an electronic companion to

calculus™ workbook

an electronic companion to

calculus™ workbook

an electronic companion to

calculus™

Keith Devlin
St. Mary's College of California

Cogito Learning Media, Inc.
New York   San Francisco

ISBN: 1-888902-52-3

COGITO END-USER AGREEMENT

PLEASE READ THE FOLLOWING CAREFULLY BEFORE USING THIS PRODUCT (INCLUDING THIS WORKBOOK AND THE ACCOMPANYING SOFTWARE AND OTHER MATERIALS) (THE "PRODUCT"). BY USING THE PRODUCT, YOU ARE AGREEING TO ACCEPT THE TERMS AND CONDITIONS OF THIS AGREEMENT. IF YOU DO NOT ACCEPT THOSE TERMS AND CONDITIONS, PLEASE RETURN THE PRODUCT (INCLUDING ANY COPIES) TO THE PLACE OF PURCHASE WITHIN 15 DAYS OF PURCHASE FOR A FULL REFUND.

LIMITATIONS ON USE

Cogito Learning Media, Inc. ("Cogito") licenses the Product to you only for use on a single computer with a single CPU. You may use the Product only on a stand-alone basis, such that the Product and the user interface and functions of the Product are accessible only to a person physically present at the location of the computer on which the Product is loaded. You may not lease the Product, display or perform the Product publicly, or allow the Product to be accessed remotely or transmitted through any network or communication link. You own the CD-ROM or other media on which the Product is recorded, but Cogito and its licensors retain all title to and ownership of the Product and reserve all rights not expressly granted to you.

You may not copy all or any portion of the Product, except as an essential step in the use of the Product as expressly authorized in the documentation included in the Product. You may transfer your license to the Product, provided that (a) you transfer all portions of the Product (including any copies) and (b) the transferee reads and agrees to be bound by the terms and conditions of this agreement. Except to the extent expressly permitted by the laws of the jurisdiction where you are located, you may not decompile, disassemble or otherwise reverse engineer the Product.

LIMITED WARRANTY

Cogito warrants that, for a period of 90 days after purchase by you (or such other period as may be expressly required by applicable law) ("Warranty Period"), (a) the Product will provide substantially the functionality described in the documentation included in the Product, if operated as specified in that documentation, and (b) the CD-ROM or other media on which the Product is recorded will be free from defects in materials and workmanship. Your sole remedy, and Cogito's sole obligation, for breach of the foregoing warranties is for Cogito to provide you with a replacement copy of the Product or, at Cogito's option, for Cogito to refund the amount paid for the Product.

EXCEPT FOR THE FOREGOING, THE PRODUCT IS PROVIDED WITHOUT WARRANTIES OF ANY KIND, EXPRESS OR IMPLIED, INCLUDING, BUT NOT LIMITED TO, ANY IMPLIED WARRANTIES OF MERCHANTABILITY OR FITNESS FOR A PARTICULAR PURPOSE. Among other things, Cogito does not warrant that the functions contained in the Product will meet your requirements, or that operation of the Product or information contained in the product will be error-free. This agreement gives you specific legal rights, and you may also have other rights which vary from state to state. Some states do not allow the exclusion of implied warranties, so the above exclusion may not apply to you. Any implied warranties will be limited to the Warranty Period, except that some states do not allow limitations on how long an implied warranty lasts, so this limitation may not apply to you.

LIMITATION OF LIABILITY

In no event will Cogito be liable for any indirect, incidental, special or consequential damages or for any lost profits, lost savings, lost revenues or lost data arising from or relating to the Product, even if Cogito has been advised of the possibility of such damages. In no event will Cogito's liability to you or any other person exceed the amount paid by you for the Product regardless of the form of the claim. Some states do not allow the exclusion or limitation of incidental or consequential damages, so the above limitation or exclusion may not apply to you. Also, this limitation will not apply to liability for death or personal injury to the extent applicable law prohibits such limitation.

GENERAL

This agreement is governed by the laws of the State of California without reference to its choice-of-law rules. This agreement is the entire agreement between you and Cogito and supersedes any other understanding or agreements. If any provision of this agreement is deemed invalid or unenforceable by any court or government agency having jurisdiction, that particular provision will be deemed modified to the extent necessary to make the provision valid and enforceable, and the remaining provisions will remain in full force and effect. Should you have any questions regarding the Product or this agreement, please contact Cogito at 1-800-WE-THINK.

1 2 3 4 5 6 7 8 9 10—RRD—00 99 98 97 Printed in U.S.A.

contents

Preface

WHAT IS THE ELECTRONIC COMPANION TO CALCULUS?

Calculus Made Easy?

Beware of any product that claims to make it easy to learn calculus. Anyone who says that is almost certainly trying to sell you something. Calculus cannot be made easy. The *Electronic Companion to Calculus*™ (consisting of a CD-ROM and this workbook) will not make it easy. It should, however, make it easi*er* to learn than if you did not have it. And it will probably make it more enjoyable as well. But it cannot take away the necessity of sitting down and putting in a lot of hard mental effort. Remember, the invention of calculus is one of the pillars of human intellectual development. It would be unreasonable to expect you to reconstruct this pillar in your own mind without blood, sweat, and tears.

The Silicon Spectrum

Where, exactly, does the *Electronic Companion to Calculus*™ fit in today's silicon spectrum of computer-aided instruction? The CD-ROM component of the *Electronic Companion* does not attempt to replace the scientific/graphing calculator or the computer algebra system. These are useful tools to have, and if you don't have one already, you should get one. (Better still, get both.) You'll need them, but not to work your way through the *Companion*. With a very small number of exceptions, the problems you will find on the *Electronic Companion* require neither a calculator nor a computer algebra system. You will, however, learn and understand calculus much faster if you do have those tools available.

The CD-ROM component of the *Electronic Companion to Calculus*™ uses the interactivity, animation, and video provided by multimedia technology to present—in a dynamic, exploratory way—the fundamental concepts of calculus and to offer additional opportunities to develop your understanding. If you are taking a first- or second-semester course in calculus (*no matter what textbook you're using*), the *Electronic Companion* can provide you with real help. Use of the *Companion* in conjunction with a graphing calculator or computer algebra system is particularly recommended.

Teach Yourself Calculus?

It *is* possible to teach yourself calculus. For some people, that is. Isaac Newton did it. So, too, did Gottfried Leibniz. They had to, since they were the two people who invented the subject, back in the mid-seventeenth century. But most people find they need an instructor. People can teach you, products can't. Products can help, but for most people they can't take the place of a real, live instructor. As the name suggests, the *Electronic Companion to Calculus*™ was developed for use as a *companion*. It was developed with the assumption that you are receiving—or at some time in the past have had—some instruction in calculus. It is not designed to teach you calculus from scratch.

Get Help

If you need help in calculus, the *Electronic Companion to Calculus*™ will be of assistance to you. But don't let it substitute for the kind of help you can get from a person, preferably a qualified mathematician. Remember, a computer can't see and understand your mistakes, and it can't keep trying different ways to help you understand. What a computer can do is provide you with an excellent way to reinforce what you have learned elsewhere, help you better understand the material you have already learned, and give you an opportunity to practice your skills.

The Key to Success

The key to success in calculus is to understand the basic underlying ideas. They are few in number—between half a dozen and a dozen, depending on what you count as a key idea. Understand them, and all of calculus is yours for the taking. The snag—for most of us—is that those underlying key ideas are totally abstract. Indeed, the success of calculus in the world of science, technology, economics and business, and other arenas of human activity is a dramatic testimony of the power of abstraction.

To find your way to the conceptual heart of calculus, you need to come to grips with abstraction. And that means you have to master the use of mathematical symbols. Just as you need to master musical notation in order to study the abstract musical patterns of the musician, so, too, you need to master mathematical notation to study the abstract mathematical patterns of calculus. To describe an abstract pattern precisely, you need an abstract notation.

The Wheel of Calculus

Mastery of abstract symbols is an essential step on the path to understanding calculus. There is no getting around that. Without the assistance of the tools of algebra, you won't be able to get to the key underlying concepts. In fact, until recently, mastery of symbols was the *only* way to get to the conceptual heart of calculus, and students not proficient in algebra were among the first to fail calculus. Today, with the arrival of sophisticated calculators and computer algebra systems, such as *Derive*, *Mathematica*, and *Maple*, new opportunities have arisen for understanding and mastering calculus.

With today's technology, we can approach the conceptual core of calculus from four perspectives: the symbolic/conceptual, the numerical, the graphical, and applications. Together, they form the wheel of calculus.

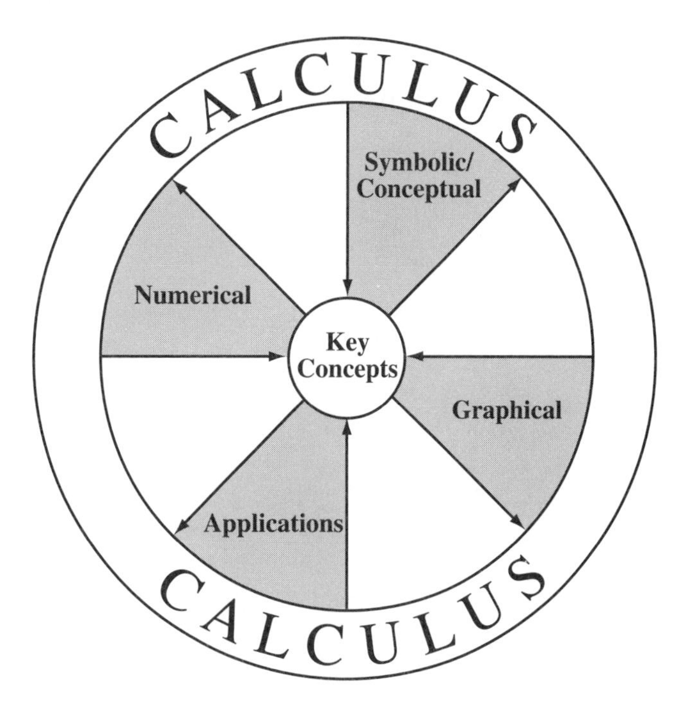

With a scientific calculator, you can explore the numerical aspects of calculus. A graphing calculator or a computer algebra system offers you a powerful graphical exploration of calculus. These same tools also let you examine realistic and extensive applications. But nothing replaces the necessity of mastering the symbolic approach. For, ultimately, it is only by way of the symbolic spoke of the wheel of calculus that you gain true access to the fundamental concepts. The *Electronic Companion to Calculus*™ is designed to help you master that all-important symbolic/conceptual aspect of calculus.

What Is The Purpose of This Workbook?

This workbook is just one of many resources you should draw upon in your effort to master calculus. It does not simply repeat material found in the CD-ROM component of the *Electronic Companion to Calculus*™. Nor is it intended to take the place of a calculus textbook. Your textbook is an indispensable reference for calculus. This workbook is designed to *supplement* your use of *any* textbook, and the CD-ROM was developed with the assumption that this workbook would offer further additional explanations and opportunities for developing problem-solving skills.

So Begin

Now. And good luck.

Keith Devlin

an electronic companion to

calculus™ workbook

dx

topic 1
Precalculus Review

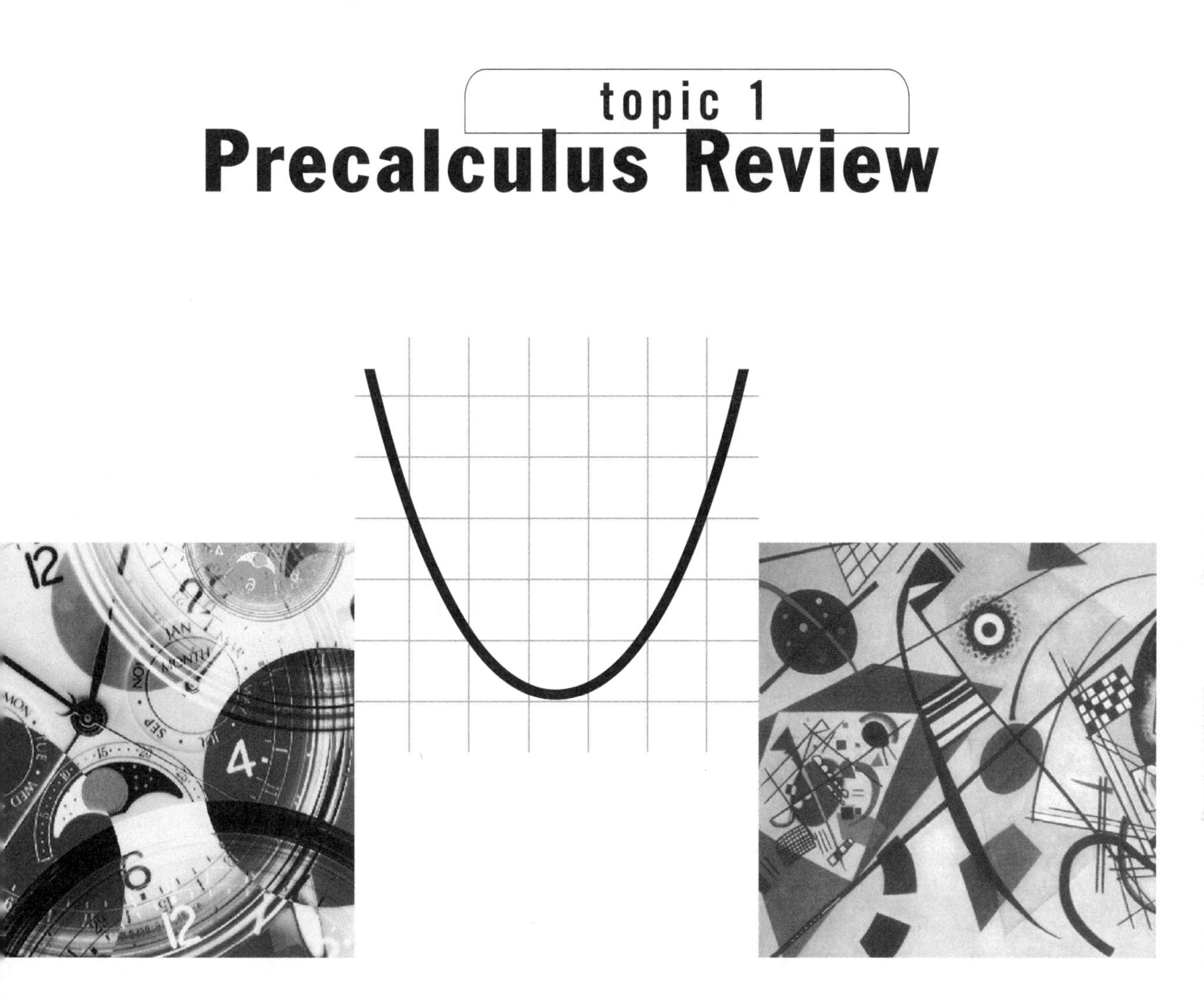

1.1 WHAT IS CALCULUS?

The subject generally known as 'calculus' really consists of two separate but related sets of mathematical techniques—the differential calculus and the integral calculus. Both branches of calculus, and the connection between them, were developed independently by Isaac Newton and Gottfried Leibniz in the seventeenth century.

Differential calculus is a collection of mathematical procedures that enable us to use the 'static' concepts of mathematics (measurement of length, classification of shape, etc.) to study continuous motion and change. The key intuition is to regard continuous motion as made up of an infinite sequence of static configurations, each one of infinitesimally small duration. This is the mathematical equivalent of a movie, where the continuous motion we see on the screen is made up of a sequence of still pictures that are projected onto the screen in rapid succession.

Integral calculus is a collection of mathematical procedures that—among other uses—enable us to study objects having a complicated (though regular) shape. Using the technique of integration, we can, for example, calculate the areas of various plane figures and the volumes of various solid objects. The key intuition is to regard the figure or object in question as made up of an infinite number of infinitely small constituents, for each of which we know how to calculate the area or volume. What makes this intuition work is the close connection it turns out to have with differentiation.

The word 'calculus' is a Latin word meaning pebble. The Romans used pebbles for counting and performing basic arithmetic, and the word grew to mean any set of procedures for performing computations of a particular kind. In both the differential and the integral calculus, the procedures deal with computations that involve infinitely small quantities—or, more precisely, the procedures enable the mathematician to skirt neatly around having to deal with infinitely small quantities.

Whereas differential and integral calculus are two well-developed branches of mathematics, there is no single branch of mathematics called 'precalculus.' The name simply refers to an assortment of mathematical topics with which you have to be familiar before you can start to learn calculus. These various topics are the subject matter of this initial chapter, and of the first section of the *Electronic Companion*.

1.2 THE REAL NUMBERS

The set of *real numbers* consists of the integers (i.e., the positive and negative whole numbers, along with zero), the rational numbers (i.e., numbers expressible as a ratio a/b of two integers, where $b \neq 0$), and the irrational numbers (i.e., numbers expressible as an infinite, non-

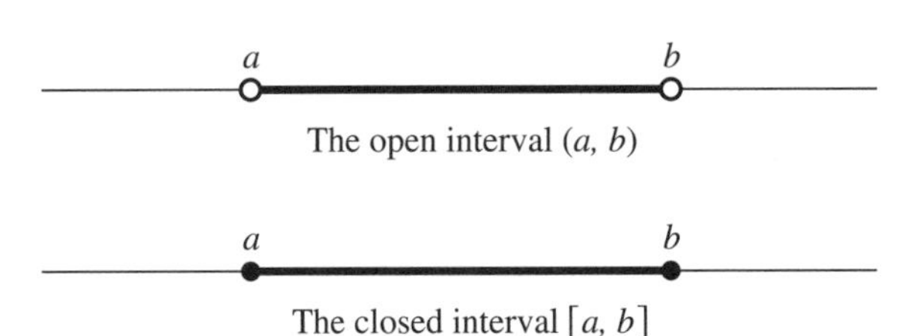

FIGURE 1.1 *Representation of intervals*

repeating decimal, such as $\pi = 3.14159\ldots$ and $\sqrt{2} = 1.41421\ldots$, all of which are non-rational).

If x and y are real numbers, $x < y$ (or $y > x$) denotes that x is less than y; $x \leq y$ (or $y \geq x$) indicates that x is either less than y or possibly equal to y.

1.2.1 Intervals

An interval is a continuous segment of the real line. If a and b are numbers, with $a < b$, the *open interval* with endpoints a and b is the set of all real numbers strictly between a and b:

$$(a, b) = \{x \mid a < x < b\}$$

The *closed interval* with endpoints a and b is the set of all real numbers between a and b inclusive:

$$[a, b] = \{x \mid a \leq x \leq b\}$$

We often use hollow dots to indicate the endpoints of an open interval, and solid dots for the endpoints, as illustrated in Figure 1.1.

Variations on the interval notation are as follows:

$$(a, b] = \{x \mid a < x \leq b\} \qquad [a, b) = \{x \mid a \leq x < b\}$$

$$(a, \infty) = \{x \mid a < x\} \qquad [a, \infty) = \{x \mid a \leq x\}$$

$$(-\infty, b) = \{x \mid x < b\} \qquad (-\infty, b] = \{x \mid x \leq b\}$$

$$(-\infty, \infty) = \text{the whole real line}$$

1.2.2 Absolute Value

The *absolute value* $|x|$ of a real number x is the distance of x from 0 measured along the real number line. It can be defined as:

$$|x| = \begin{cases} x, & \text{if } x \geq 0 \\ -x, & \text{if } x < 0 \end{cases}$$

For example, $|5.7| = 5.7$ and $|-20.83| = 20.83$.
Basic properties of absolute value:

1. $|x| \geq 0$
2. $|-x| = |x|$
3. $-|x| \leq x \leq |x|$
4. $|x - y| = |y - x|$
5. $|x + y| \leq |x| + |y|$ (the triangle inequality)

1.2.3 Inequalities

You can solve an inequality in much the same way you can solve an equation. Whereas the solution to an equation is generally one or more real numbers, the solution to an inequality is typically one or more intervals.

You can add the same quantity to both sides of an inequality without affecting the validity of the inequality. Likewise, you can subtract the same quantity from both sides. For example, if

$$x \leq y$$

then, for any number a,

$$x + a \leq y + a \text{ and } x - a \leq y - a$$

If you multiply both sides of a valid inequality by a positive quantity, it remains valid. If you multiply both sides of a valid inequality by a negative quantity, the inequality reverses direction. For example, if

$$x < y$$

then, for any positive number a,

$$ax < ay \text{ but } (-a)x > (-a)y$$

Since division by a non-zero number a is the same as multiplication by $1/a$, the last remark also applies to division.

An inequality of the form

$$|E(x)| \leq M$$

(where $E(x)$ is some expression involving x) that involves an absolute value can be replaced by the pair of inequalities

$$-M \leq E(x) \leq M$$

that does not involve an absolute value.

When you are faced with solving an inequality, you will often find it useful to split it into two or more cases, depending on the signs of the various constituent expressions.

1.2.4 Examples

Example 1

Solve the inequality

$$x^2 + 3x - 4 > 0$$

Solution

Factor the quadratic, to give

$$(x - 1)(x + 4) > 0$$

There are two cases to consider.

Case 1. $x - 1 > 0$ and $x + 4 > 0$.
That is, $x > 1$ and $x > -4$. In this case, both inequalities are satisfied if $x > 1$.

Case 2. $x - 1 < 0$ and $x + 4 < 0$.
That is, $x < 1$ and $x < -4$. In this case, both inequalities are satisfied if $x < -4$.

Thus, the complete solution is the pair of intervals $(-\infty, -4)$, $(1, \infty)$.

Example 2

Solve the inequality

$$\frac{3x - 4}{x + 2} < 5$$

Solution

The idea is to split it into two cases, depending on the sign of the denominator: $x + 2 < 0$ or $x + 2 > 0$.

Case 1. $x > -2$.
Multiplying through by the denominator, which is positive, the inequality becomes

$$3x - 4 < 5(x + 2)$$

which simplifies to

$$3x - 4 < 5x + 10$$

Hence, $-14 < 2x$, giving the solution $x > -7$. This is automatically satisfied in this case, since $x > -2$. Thus the solution interval in this case is $(-2, \infty)$.

Case 2. $x < -2$.
Multiplying through by the denominator, which is negative, the inequality becomes

$$3x - 4 > 5(x + 2)$$

which simplifies to

$$3x - 4 > 5x + 10$$

Hence, $-14 > 2x$, giving the solution $x < -7$. Thus, in this case, the solution interval is $(-\infty, -7)$.

Thus the entire solution to the inequality is $(-\infty, -7), (-2, \infty)$.

✎ Example 3

Solve the inequality

$$\frac{3x + 4}{x - 3} < 2$$

Solution

The idea is to split it into cases, depending on the sign of the denominator: $x - 3 > 0$ or $x - 3 < 0$.

Case 1. $x > 3$. Since $x - 3 > 0$, the inequality becomes

$$3x + 4 < 2x - 6$$

which simplifies to $x < -10$. This is incompatible with the assumption for Case 1 (namely $x > 3$), so there is no solution in this case.

Case 2. $x < 3$. The inequality becomes

$$3x + 4 > 2x - 6$$

which simplifies to $x > -10$. Now, in Case 2 we also have the assumption $x < 3$. So the overall solution set is the interval $(-10, 3)$.

✎ Example 4

Solve the inequality

$$\left|\frac{1}{x} - 2\right| < 5$$

Solution

The idea is to rewrite the inequality as

$$-5 < \frac{1}{x} - 2 < 5$$

Adding 2 to each term, this becomes

$$-3 < \frac{1}{x} < 7$$

Now split into two cases depending on the sign of x.

Case 1. $x > 0$.
Multiplying through by x, the inequality becomes

$$-3x < 1 < 7x$$

Since $x > 0$ in this case, $-3x < 1$ is automatically true. Thus the inequality reduces to $x > \frac{1}{7}$.

Case 2. $x < 0$.
Multiplying through by x, the inequality becomes

$$-3x > 1 > 7x$$

The inequality $1 > 7x$ is automatically true in this case, since $x < 0$. Thus the inequality reduces to $x < -\frac{1}{3}$, and thus the entire solution set is $(-\infty, -\frac{1}{3}), (\frac{1}{7}, \infty)$.

1.2.5 Problems

1. Solve the following inequalities:

(a) $|5x - 1| \geq 2$

(b) $x^2 > x + 2$

(c) $(x + 1)(x + 2)(x + 3) > 0$

(d) $\dfrac{x}{x+1} < 1$

(e) $\left|\dfrac{3x+2}{x+3}\right| > 3$

2. Using a graphing calculator, graph each of the functions in examples 1, 2, and 3 in the above Examples section, and see how the behavior of the graph relates to the solution of the inequality.

1.3 CARTESIAN COORDINATES

Relative to a pair of Cartesian axes, we can specify any point P in the plane by a pair of real numbers: its *x-coordinate* (the distance of P from the y-axis, measured in the x-direction) and its *y-coordinate* (the distance of P from the x-axis, measured in the y-direction).

A negative number indicates distance measured in the negative direction—to the left along the x-axis, down along the y-axis.

If a point P has x-coordinate a and y-coordinate b, we use the notation (a, b) to indicate the *Cartesian coordinates* of P.

See Figure 1.2 (p. 8) for examples.

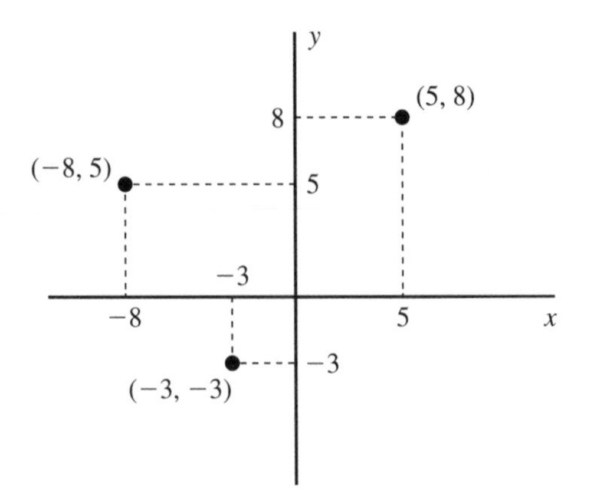

FIGURE 1.2 *Cartesian coordinates*

1.3.1 Straight Lines

If a straight line passes through the points $A = (x_1, y_1)$ and $B = (x_2, y_2)$, its *slope* m is (see Figure 1.3):

$$m = \frac{\text{change in } y}{\text{change in } x} = \frac{y_2 - y_1}{x_2 - x_1}$$

The *Cartesian equation* of the line is the algebraic equation satisfied by the coordinates (x, y) of all points on the line. It is:

$$y - y_1 = m(x - x_1)$$

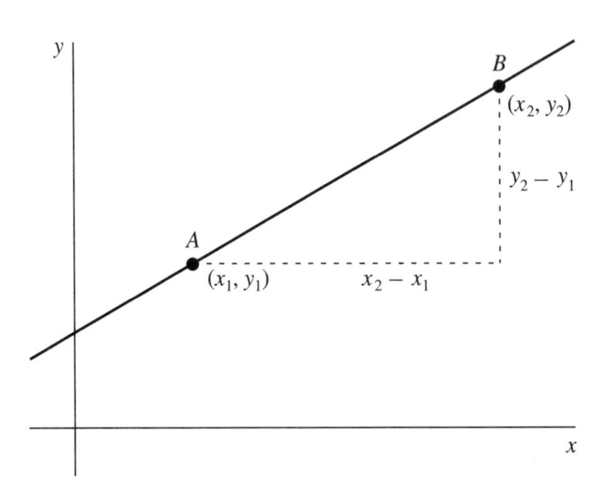

FIGURE 1.3 *Slope of a straight line*

In any specific case, where numerical values of x_1, x_2, y_1, y_2 (and hence also m) are known, this may be manipulated into the form

$$y = mx + c$$

The value c is called the *y-intercept* of the line. It is the value of the equation when $x = 0$. If m is positive, the line slopes up to the right. If m is negative, the line slopes down to the right. If $m = 0$, the line is horizontal.

1.3.2 Distance

The distance d between two points $P = (x_1, y_1)$ and $Q = (x_2, y_2)$ measured along the straight line joining the two points is given by the formula

$$d = \sqrt{|x_2 - x_1|^2 + |y_2 - y_1|^2}$$

This formula comes from Pythagoras' Theorem. If you complete right triangle PQR as shown in Figure 1.4, then the length of the side PR is $|x_2 - x_1|$ and the length of the side RQ is $|y_2 - y_1|$.

Since squaring any real number gives a positive answer, the absolute value signs may be omitted from the distance formula:

$$d = \sqrt{(x_2 - x_1)^2 + (y_2 - y_1)^2}$$

1.3.3 Examples

Example 1

Show that the triangle with vertices $P(-5, 6)$, $Q(2, 3)$, $R(5, 10)$ is a right triangle.

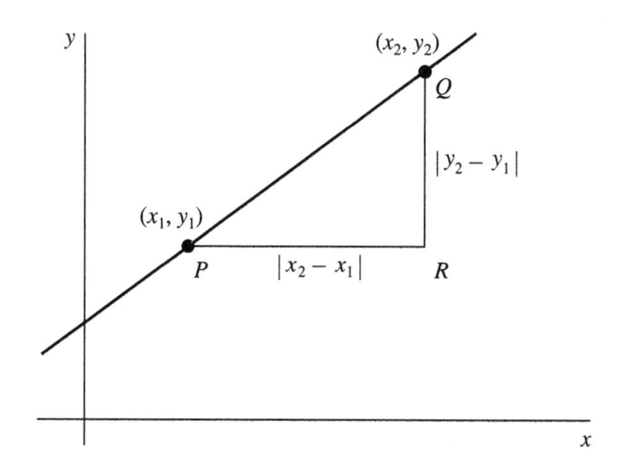

FIGURE 1.4 *Distance between two points*

Solution

Using the Pythagorean formula for the distance between two points,

$$\begin{aligned} PQ &= \sqrt{(2+5)^2+(3-6)^2} = \sqrt{49+9} = \sqrt{58} \\ PR &= \sqrt{(5+5)^2+(10-6)^2} = \sqrt{100+16} = \sqrt{116} \\ QR &= \sqrt{(5-2)^2+(10-3)^2} = \sqrt{9+49} = \sqrt{58} \end{aligned}$$

Thus,

$$PQ^2 + QR^2 = 58 + 58 = 116 = PR^2$$

So, by the (valid) inverse to Pythagoras' theorem, the triangle is right.

✎ ***Example 2***

Show that the points $A(1,5)$, $B(2,7)$, $C(3,9)$ are collinear.

Solution

It is enough to show that the slope of AB equals the slope of AC. (Do you see why?)

$$\text{Slope of } AB = \frac{7-5}{2-1} = \frac{2}{1} = 2$$

$$\text{Slope of } AC = \frac{9-5}{3-1} = \frac{4}{2} = 2$$

Done.

✎ ***Example 3***

Find the equation of the line through the points $A(3,7)$ and $B(6,11)$.

Solution

Applying the standard formula, the equation is

$$\begin{aligned} y-7 &= \left(\frac{11-7}{6-3}\right)(x-3) \\ y-7 &= \frac{4}{3}(x-3) \\ 3(y-7) &= 4(x-3) \\ 3y-21 &= 4x-12 \\ 3y &= 4x+9 \end{aligned}$$

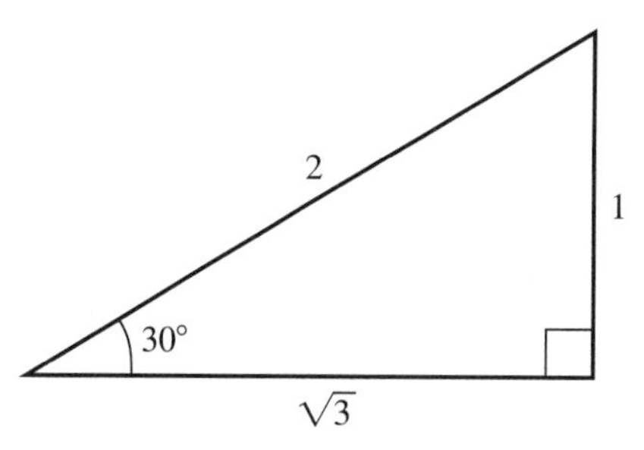

FIGURE 1.5 *Solution to Example 4*

✎ ***Example 4***

What is the equation of the line through the point $(1, 2)$ that makes an angle of 30° with the x-axis?

Solution

The slope of the line at 30° to the x-axis is $\frac{1}{\sqrt{3}}$. (A right triangle with height 1 and hypotenuse 2 has a base angle of 30° and a base-length of $\sqrt{3}$. See Figure 1.5.) So the equation of the line is

$$y = \frac{1}{\sqrt{3}}x + c$$

for some constant c. Since the point $(1, 2)$ lies on the line, we must have

$$2 = \frac{1}{\sqrt{3}}1 + c$$

so

$$c = 2 - \frac{1}{\sqrt{3}}$$

Hence the equation of the line is

$$y = \frac{1}{\sqrt{3}}x + \left[2 - \frac{1}{\sqrt{3}}\right]$$

1.3.4 Problems

1. Is the triangle with vertices $P(1, 2)$, $Q(3, 5)$, $R\ (2, 6)$ isosceles?
2. Is the triangle with vertices $P(1, 2)$, $Q(-1, 5)$, $R(3, 5)$ isosceles?
3. What are the coordinates of the point M mid-way between $A(-5, 11)$ and $B(8, -2)$?
4. What is the equation of the line through the points $(-6, -15)$ and $(8, 4)$?

5. What is the equation of the line with slope 60° that cuts the y-axis at $y = 7$?

6. Let L_1 be a straight line with slope m_1 and let L_2 be a straight line with slope m_2. Prove that L_1 and L_2 are perpendicular if and only if $m_1 m_2 = -1$. (Hint: Use similar triangles.)

Use the above result to show that the lines

$$y = 3x + 5 \text{ and } 3y = -x + 11$$

intersect at a right angle.

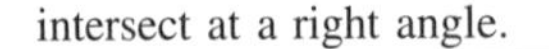

1.4 FUNCTIONS AND EQUATIONS

A function is a rule that to each element x of a set X associates a unique element y of another set Y. The set X is called the *domain* of the function; the set Y is sometimes called the *co-domain*. If f denotes the function, it is common to write $f(x)$ for the unique element y that f associates with x. Any element x of X to which f is applied is called an *argument* of f; the element $f(x)$ is called the *value* of f for the argument x. The set

$$\{f(x) \mid x \in X\}$$

of all values of f is called the *range* of f.

For example, the equation

$$y = x^2 + 5x - 1$$

determines a function f from the set $\mathcal{R}$ of real numbers to the (same) set $\mathcal{R}$ of real numbers. For the argument 3, the function f has the value

$$f(3) = (3)^2 + 5(3) - 1 = 23$$

Here are some further examples of functions:

- $f(x) = \sqrt{x}$, with domain $[0, \infty)$. The range is $[0, \infty)$.
- $f(x) = -\sqrt{x}$, with domain $[0, \infty)$. The range is $(-\infty, 0]$.
- $f(x) =$ the least prime number p such that $p \geq x$,
  with domain the entire real line. The range is the set of prime numbers.
- $f(C) =$ the capital city of C,
  where the domain is the set of all countries in the world. The range is the set of all capital cities.
- $f(a) =$ the social security number of a,
  where the domain is the set of all US citizens. The range is the set of all current social security numbers.

1.4.1 Inverses

For some functions f, given any y in the range of f, you can find a unique x in the domain that f sends to y. This defines a function that 'goes the opposite way' to f and has the effect of 'undoing' f. If $x = g(y)$ is such a function, then g is said to be an *inverse* function to f.

The relationship between a function f and an inverse function g is given by the equivalence:

$$f(x) = y \Leftrightarrow x = g(y)$$

where $\Leftrightarrow$ is read as "is equivalent to."

A function $f(x)$ can have at most one inverse function. If $f(x)$ has an inverse function, the domain of the inverse function will be the range of f. The unique inverse function of f is usually denoted by f^{-1}.

For example, the function $y = x^2$ from the set $\mathcal{R}^+$ of non-negative reals to the same set $\mathcal{R}^+$ has the inverse $x = \sqrt{y}$ (also with domain $\mathcal{R}^+$):

$$y = x^2 \Leftrightarrow x = \sqrt{y}$$

Note that, in the above example, it is important that the domain of the first function is $\mathcal{R}^+$ and not the entire real line. If we were to take the function $y = x^2$ with domain $\mathcal{R}$, then there would not be an inverse function, since for each non-zero y in $\mathcal{R}^+$ there is not one but *two* numbers x that the function sends to y, namely the positive and negative square roots of y.

To find the inverse of a function given in the form

$$y = \text{ a formula involving } x$$

you try to solve the equation to find x in terms of y.

For example, the function

$$y = 3x + 4$$

has inverse

$$x = \frac{1}{3}y - \frac{4}{3}$$

The duplication problem described a moment ago that prevents the function $y = x^2$ from having an inverse is the only reason why a function does not have an inverse. For any function f, if no two numbers in the domain of f give the same value of f, then an inverse *does* exist.

A function f with the property that different arguments a, b produce *different* values $f(a)$, $f(b)$ is called a *one-to-one* function. If f is a one-to-one function with domain A and range B, then f has an inverse function f^{-1} with domain B and range A.

1.4.2 Examples

✎ ***Example 1***

For the function

$$f(x) = \frac{x-3}{x^2+1}$$

find **(a)** $f(0)$, **(b)** $f(2)$, **(c)** $f(3)$, **(d)** $f(a)$, **(e)** $f\left(\frac{1}{x}\right)$, **(f)** $f(x+h)$.

Solution

(a) $f(0) = \frac{0-3}{0+1} = -3.$

(b) $f(2) = \frac{2-3}{4+1} = -\frac{1}{5}.$

(c) $f(3) = \frac{3-3}{9+1} = 0.$

(d) $f(a) = \frac{a-3}{a^2+1}.$

(e) $f\left(\frac{1}{x}\right) = \frac{\frac{1}{x}-3}{\left(\frac{1}{x}\right)^2+1} = \frac{x-3x^2}{1+x^2}.$

(f) $f(x+h) = \frac{(x+h)-3}{(x+h)^2+1} = \frac{x+h-3}{x^2+2xh+h^2+1}.$

✎ ***Example 2***

Find the domains of the following functions:

(a) $f(x) = \sqrt{x^2-1}$

(b) $f(x) = \sqrt{4-x^2}$

(c) $f(x) = \frac{1}{\sqrt{4-x^2}}$

(d) $f(x) = \frac{1}{x-9}$

(e) $f(x) = \frac{x}{x^2+1}$

Solution

(a) We need $x^2-1 \geq 0$, i.e., $x^2 \geq 1$, so the domain is $(-\infty, -1]$, $[1, \infty)$.

(b) We need $4 - x^2 \geq 0$, i.e., $x^2 \leq 4$, so the domain is $[-2, 2]$.

(c) The function is not defined if $x = \pm 2$. So, from **(b)**, the domain is $(-2, 2)$.

(d) The function is defined for all x except $x = 9$, so the domain is $(-\infty, 9)$, $(9, \infty)$.

(e) The domain is the entire real line, since $x^2 + 1 \neq 0$ for all x.

✎ ***Example 3*** Show that the following functions are one-to-one and give their inverses:

(a) $f(x) = 5x + 4$

(b) $f(x) = x^3$

Solution

(a) $x_1 \neq x_2$ implies $5x_1 \neq 5x_2$ implies $5x_1 + 4 \neq 5x_2 + 4$, so f is one-to-one. To find the inverse, we solve the equation

$$y = 5x + 4$$

for x. Write the equation as

$$5x = y - 4$$

Thus,

$$x = \tfrac{1}{5}y - \tfrac{4}{5}$$

Hence, swapping x and y,

$$f^{-1}(x) = \tfrac{1}{5}x - \tfrac{4}{5}$$

(b) $x_1 < x_2$ implies $x_1^3 < x_2^3$, so f is one-to-one. The inverse is immediate:

$$f^{-1}(x) = x^{1/3}$$

1.4.3 Problems

1. Find the domains of the following functions:

   (a) $f(x) = \sqrt{x^2 - 9}$

   (b) $f(x) = \sqrt{1 - x^2}$

(c) $f(x) = \dfrac{1}{\sqrt{1-x^2}}$

(d) $f(x) = \dfrac{x}{x-1}$

(e) $f(x) = 6$

2. Find the inverses to the following functions:

(a) $f(x) = 6x + 11$

(b) $f(x) = \dfrac{x}{x+1}$, where the domain is $[0, \infty)$.

1.5 TRIGONOMETRY

1.5.1 Radian Measure

In calculus, we measure angles in units called *radians*. We do this because the resulting formulas are much simpler. To obtain the radian measure of the angle θ ('theta') shown in Figure 1.6, first draw an arc AB of a circle whose center is at the point O. If s is the length of the arc and r is the radius of the circle, the radian measure of θ is the ratio s/r. (This ratio is the same whatever the value of r, since all circles are similar; the arc-length s varies directly with r.) Table 1.1 provides useful comparisons between radian measure and degrees.

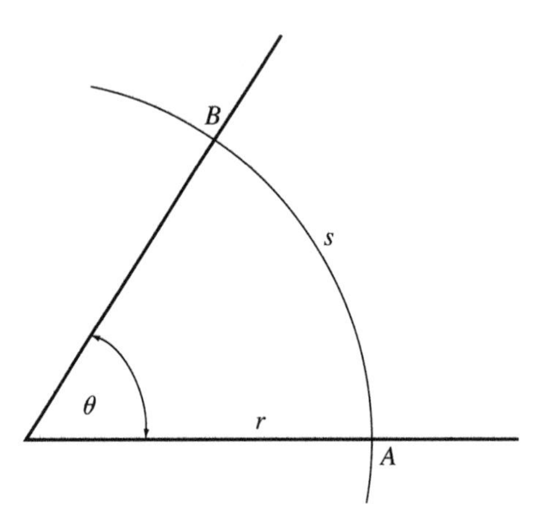

FIGURE 1.6 *Radian measure*

TABLE 1.1	Radians and Degrees
1 rad $\approx$ 57.3°	1° $\approx$ 0.0175 rad
30° $= \pi/6$ rad	45° $= \pi/4$ rad
60° $= \pi/3$ rad	90° $= \pi/2$ rad
180° $= \pi$ rad	360° $= 2\pi$ rad

1.5.2 Trig Functions

For an angle θ between 0 and $\pi/2$ radians, the trigonometric (or 'trig') functions can be defined in terms of the ratios of the sides of a right triangle having θ as one of its angles, as shown in Figure 1.7. Here are the definitions for the three common trig functions sine, cosine, and tangent, and their reciprocals cosecant, secant, and cotangent:

$$\sin\theta = \frac{\text{opposite}}{\text{hypotenuse}} = \frac{BC}{AB} \qquad \csc\theta = \frac{1}{\sin\theta}$$

$$\cos\theta = \frac{\text{adjacent}}{\text{hypotenuse}} = \frac{AC}{AB} \qquad \sec\theta = \frac{1}{\cos\theta}$$

$$\tan\theta = \frac{\text{opposite}}{\text{adjacent}} = \frac{BC}{AC} \qquad \cot\theta = \frac{1}{\tan\theta}$$

Notice that, for any θ:

$$\tan\theta = \frac{\sin\theta}{\cos\theta} \qquad \cot\theta = \frac{\cos\theta}{\sin\theta}$$

For an arbitrary angle θ, not simply one between 0 and $\pi/2$, the values of $\sin\theta$ and $\cos\theta$ are given by the Cartesian coordinates of points on the unit circle, whose center is at the origin. If the radius from the origin to the point P on the circle subtends an angle θ with the x-axis, then the coordinates of P are given by

$$x = \cos\theta, \quad y = \sin\theta$$

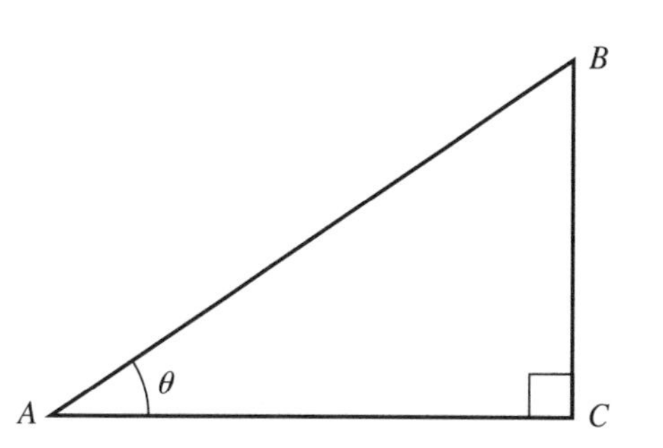

FIGURE 1.7 *Definition of the trigonometric functions*

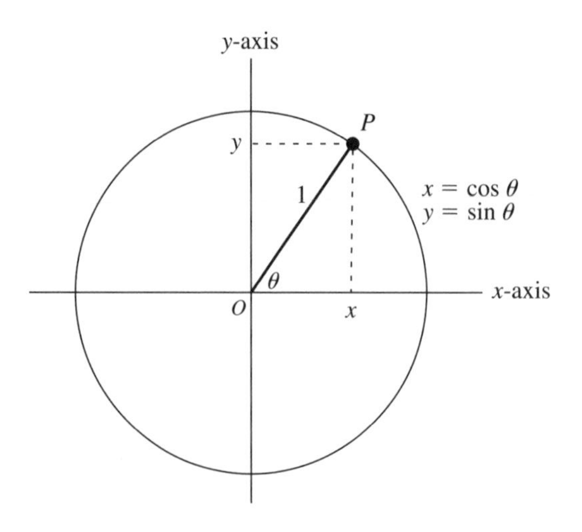

FIGURE 1.8 *Circular definition of the trigonometric functions*

Here, θ can be any real value. If θ is positive, the angle is measured counterclockwise; a negative value for θ indicates a clockwise measurement. See Figure 1.8.

When we measure angles, an angle of 2π is equivalent to an angle of 0. Thus, the trig functions sine and cosine are *periodic* functions, with a *period* of 2π: the value of each function is the same for θ and $\theta + 2\pi$. Figure 1.9 illustrates a part of their graphs.

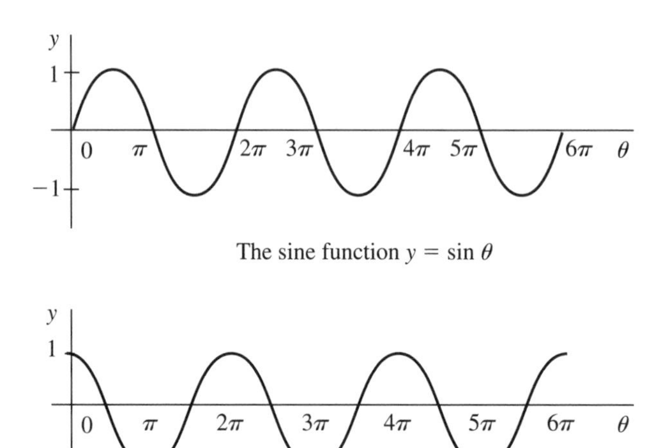

FIGURE 1.9 *The sine and cosine functions*

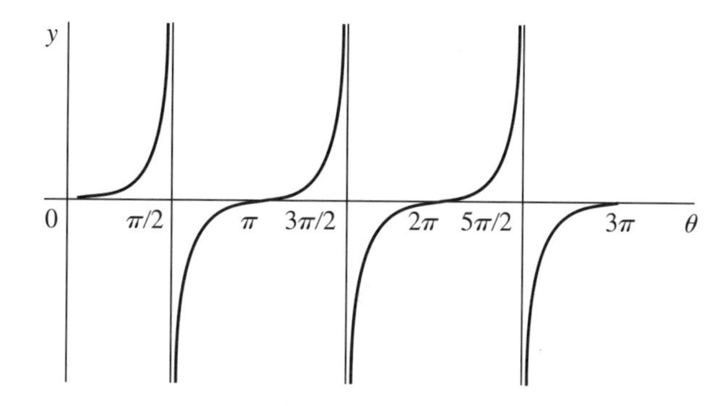

FIGURE 1.10 *The tangent function:* $y = \tan\theta$

Since the other trig functions can be defined as reciprocals or ratios of sine and cosine, it follows that they too are periodic (with a period of 2π or less). For instance, the tangent function is periodic with period π. Part of the graph of the tangent function is given in Figure 1.10.

Using the triangles shown in Figure 1.11, we can compute the trig values in Table 1.2 on p. 20. Values of the other trig functions at the same angles can be deduced from the values in the table.

1.5.3 Trig Identities

The following trig identities are used frequently in calculus:

$$\sin(-\theta) = -\sin(\theta) \qquad \cos(-\theta) = \cos(\theta) \qquad \tan(-\theta) = -\tan(\theta)$$

$$\sin(\tfrac{\pi}{2} - X) = \cos X \qquad \cos(\tfrac{\pi}{2} - X) = \sin X$$

$$\sin^2 X + \cos^2 X = 1$$

$$\sec^2 X = 1 + \tan^2 X \qquad \csc^2 X = 1 + \cot^2 X$$

$$\sin(2X) = 2 \sin X \cos X$$

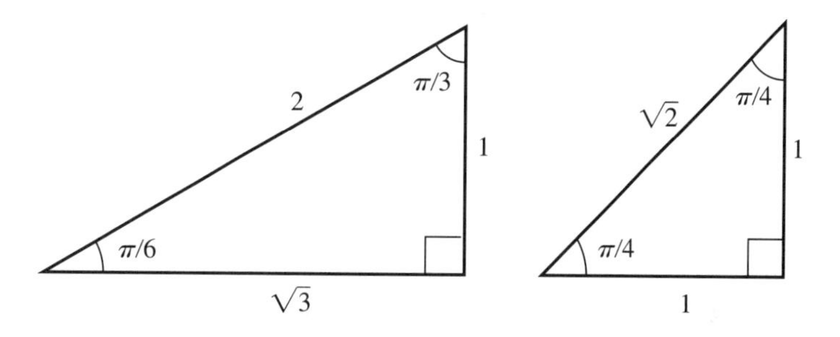

FIGURE 1.11 *Special angles*

TABLE 1.2 Special Trig Values

X	$\sin X$	$\cos X$	$\tan X$
0	0	1	0
$\frac{\pi}{6}$	$\frac{1}{2}$	$\frac{\sqrt{3}}{2}$	$\frac{1}{\sqrt{3}}$
$\frac{\pi}{4}$	$\frac{\sqrt{2}}{2}$	$\frac{\sqrt{2}}{2}$	1
$\frac{\pi}{3}$	$\frac{\sqrt{3}}{2}$	$\frac{1}{2}$	$\sqrt{3}$
$\frac{\pi}{2}$	1	0	undefined
π	0	-1	0

$$\cos(2X) = \cos^2 X - \sin^2 X = 2\cos^2 X - 1 = 1 - 2\sin^2 X$$
$$\sin(X+Y) = \sin X \cos Y + \cos X \sin Y$$
$$\sin(X-Y) = \sin X \cos Y - \cos X \sin Y$$
$$\cos(X+Y) = \cos X \cos Y - \sin X \sin Y$$
$$\cos(X-Y) = \cos X \cos Y + \sin X \sin Y$$
$$\tan(X+Y) = (\tan X + \tan Y)/(1 - \tan X \tan Y)$$
$$\tan(X-Y) = (\tan X - \tan Y)/(1 + \tan X \tan Y)$$

1.5.4 Examples

Example 1

Evaluate: $\sin(3\pi)$, $\cos\left(\frac{9\pi}{4}\right)$, $\sin\left(\frac{11\pi}{6}\right)$

Solution

$$\begin{aligned} \sin(3\pi) &= \sin(2\pi + \pi) = \sin(\pi) = 0 \\ \cos\left(\tfrac{9\pi}{4}\right) &= \cos\left(2\pi + \tfrac{\pi}{4}\right) = \cos\left(\tfrac{\pi}{4}\right) = \tfrac{1}{\sqrt{2}} \\ \sin\left(\tfrac{11\pi}{6}\right) &= \sin\left(2\pi - \tfrac{\pi}{6}\right) = \sin\left(-\tfrac{\pi}{6}\right) = -\tfrac{1}{2} \end{aligned}$$

Example 2

Starting with the identity

$$\sin(X-Y) = \sin X \cos Y - \cos X \sin Y$$

prove the identity

$$\cos(X+Y) = \cos X \cos Y - \sin X \sin Y$$

Solution

$$\begin{aligned}
\cos(X+Y) &= \sin(\tfrac{\pi}{2} - (X+Y)) \\
&= \sin((\tfrac{\pi}{2} - X) - Y) \\
&= \sin(\tfrac{\pi}{2} - X)\cos Y - \cos(\tfrac{\pi}{2} - X)\sin Y \\
&= \cos X \cos Y - \sin X \sin Y
\end{aligned}$$

✎ ***Example 3***

Starting with the standard identities

$$\sin(A+B) = \sin A \cos B + \cos A \sin B$$
$$\cos(A+B) = \cos A \cos B - \sin A \sin B$$

show that

$$\tan(A+B) = \frac{\tan A + \tan B}{1 - \tan A \tan B}$$

Solution

Recall the relationship between tangent, sine, and cosine:

$$\tan A = \frac{\sin A}{\cos A}$$

Using the above identity and the equivalent one for B,

$$\begin{aligned}
\frac{\tan A + \tan B}{1 - \tan A \tan B} &= \frac{(\sin A/\cos A) + (\sin B/\cos B)}{1 - (\sin A/\cos A)(\sin B/\cos B)} \\
&= \frac{\sin A \cos B + \cos A \sin B}{\cos A \cos B - \sin A \sin B} \\
&= \frac{\sin(A+B)}{\cos(A+B)} \\
&= \tan(A+B)
\end{aligned}$$

✎ ***Example 4***

Show that

$$\sec^2 A = 1 + \tan^2 A$$

Solution

Expressing $\tan A$ in terms of $\sin A$ and $\cos A$,

$$\begin{aligned} 1+\tan^2 A &= 1+\frac{\sin^2 A}{\cos^2 A} \\ &= \frac{\cos^2 A+\sin^2 A}{\cos^2 A} \\ &= \frac{1}{\cos^2 A} \quad (\text{since} \ \ \cos^2 A+\sin^2 A=1) \\ &= \sec^2 A \end{aligned}$$

as required.

1.5.5 Problems

1. Evaluate:

$$\sin(21\pi) \qquad \cos(-11\pi) \qquad \tan(9\pi)$$

$$\sin(\tfrac{13\pi}{6}) \qquad \cos(\tfrac{19\pi}{3}) \qquad \tan(\tfrac{9\pi}{4})$$

2. Show that

$$\tan\theta = \frac{\sin 2\theta}{1+\cos 2\theta}$$

3. Show that

$$(\sin\theta+\cos\theta)^2 = 1+\sin 2\theta$$

1.6 EXPONENTIALS AND LOGS

1.6.1 Exponentiation

For any real number a and for any integer $n \geq 0$, the number a^n is defined as follows:

$$\begin{aligned} a^0 &= 1 \\ a^1 &= a \\ a^2 &= a \times a \\ a^3 &= a \times a \times a \end{aligned}$$

etc.

In the case where $a > 0$, we define $a^{1/n}$ to be the positive nth root of a, $\sqrt[n]{a}$. We then define a^q for any positive rational number $q = m/n$ (where m, n are positive integers) by $a^q = (a^{1/n})^m$.

We define a^r for negative rational numbers $r = -q$ (q a positive rational) by setting $a^{-q} = \dfrac{1}{a^q}$.

When you compute a^n from a you are 'raising a to the power n.' The quantity a^n is called the nth *power* of a.

There is a general function, called the *exponential function*, which, for any positive real number a, gives a value a^x for any real number x (not just integers or rationals), and which gives the n th power of a in the case where x is an integer or a rational number n. In the expression a^x, we call x the *exponent* of a.

The formal definition of the exponential function is beyond the scope of this treatment. For any positive real number a, the function a^x has the following properties:

1. $a^0 = 1,\ a^1 = a$
2. $a^x \times a^y = a^{x+y}$
3. $(a^x)^y = a^{xy}$
4. $a^{-x} = 1/a^x$
5. If n is a positive integer, a^n is the n th power of a
6. If n is a positive integer, $a^{1/n} = \sqrt[n]{a}$, the n th root of a

Graphs of the exponential function a^x for some different values of a are shown in Figure 1.12. Notice the difference between the two cases $0 < a < 1$ and $a > 1$.

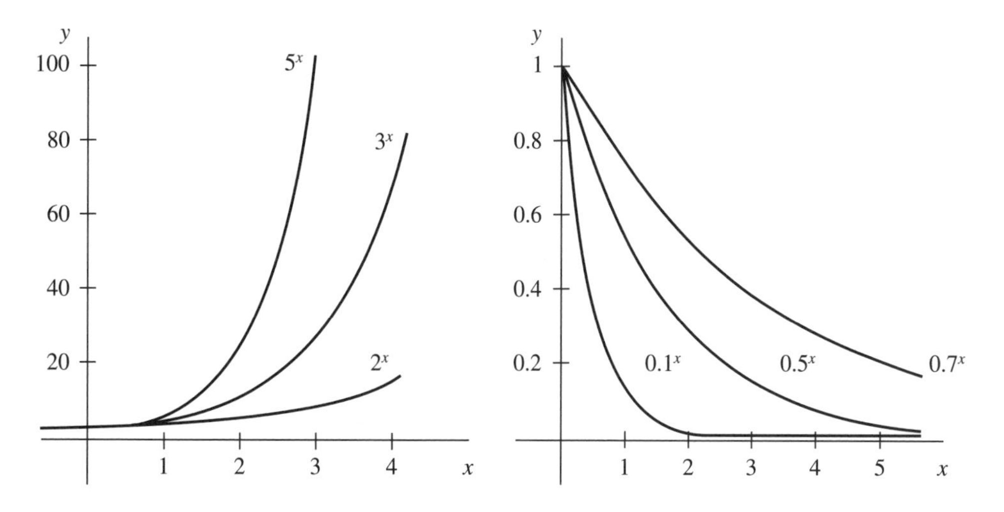

FIGURE 1.12 *The exponential function* $y = a^x$

1.6.2 Logarithms

For any positive real number a, the *base-a logarithm* function, $\log_a x$, is defined to be the inverse to the exponential function a^x. That is,

$$y = \log_a x \iff x = a^y$$

Let a be a fixed positive real number. Then the function $\log_a$ has the following properties:

1. $\log_a(xy) = \log_a x + \log_a y$
2. $\log_a(x/y) = \log_a x - \log_a y$
3. $\log_a(x^b) = b \log_a x \qquad (b \text{ any real number})$
4. $\log_a(1/x) = -\log_a x$
5. $\log_a a = 1, \quad \log_a 1 = 0$

In calculus, logarithms to base e are particularly important, where e is the special mathematical constant whose definition is given in Section 3.4. The number e is irrational; it is approximately equal to 2.71828. Logarithms to base e are generally written as ln; thus

$$\ln x = \log_e x$$

It follows that

$$y = \ln x \text{ iff } x = e^y$$

Hence,

$$\ln(e^x) = e^{\ln x} = x$$

Logarithms to base e are called *natural logarithms.*

1.6.3 Examples

Example 1

Let e be any positive real number. (The letter e is used here because the most important instances of these examples is where the base is the special constant $e = 2.71828\ldots$. But the examples are general results that are valid for any base.) Evaluate:

$$e^{5\log_e x},\ e^{-\log_e x},\ \log_e e^{-x},\ \log_e\left(\frac{x}{e^x}\right),\ e^{\log_e(x+1)}$$

Solution

$$e^{5\log_e x} = \left(e^{\log_e x}\right)^5 = x^5$$

$$e^{-\log_e x} = \left(e^{\log_e x}\right)^{-1} = x^{-1} = \frac{1}{x}$$

$$\log_e(e^{-x}) = -x \log_e e = -x$$

$$\log_e\left(\frac{x}{e^x}\right) = \log_e x - \log_e e^x = \log_e x - x$$

$$e^{\log_e(x+1)} = x + 1$$

✎ ***Example 2***

Show that the graph of $y = \ln x$ is the reflection of the graph of $y = e^x$ in the line $y = x$. (More generally, the graph of $y = \log_m x$ is the reflection of the graph of $y = m^x$ in the line $y = x$ for any positive real number m.) [See Problem 2 below.]

Solution

A point (a, b) is on the graph of $y = e^x \iff$

$$\begin{aligned} & b = e^a \\ \iff & \ln b = a \\ \iff & (b, a) \text{ is on the graph of } y = \ln x \end{aligned}$$

But the points (a, b) and (b, a) are symmetric with respect to the line $y = x$.

1.6.4 Problems

1. Let e be any positive real number. (A useful special case is when e is the special constant $e = 2.718\ldots.$) Evaluate

$$e^{\log_e 2}, \quad \log_e e^3, \quad \log_e \sqrt{e}, \quad \log_e(e^{x+1}), \quad \log_e\left(\frac{x+1}{e^2}\right)$$

2. Use a graphing calculator or a computer algebra system to draw graphs of the three functions $y = \ln x$, $y = e^x$, $y = x$ for different ranges. Start by letting x and y range from 0 to 0.1, from 0.1 to 0.5, and from 0.5 to 1, and then choose your own ranges.

3. Use a graphing calculator to check the following identities graphically.

   (1) $\ln(xy) = \ln x + \ln y \quad (x > 0, \quad y > 0)$

   (2) $\ln(x/y) = \ln x - \ln y \quad (x > 0, \quad y > 0)$

   (3) $\ln(x^b) = b \ln x \quad$ (b any real number, $x > 0$)

   (4) $\ln(1/x) = -\ln x \quad (x > 0)$

   In parts 1 and 2, choose various constant values for y.

dx

topic 2
Limits and Continuity

The related concepts of limits and continuity are fundamental to calculus. The idea of a limit comes up in two forms, limits of sequences and limits of functions. The definitions for the two cases are similar, but neither depends on the other. The sequences case is arguably a bit easier to understand, which is one reason for covering it first, as I do here. Another reason for introducing sequences first is that most present-day textbooks do it the other way round, so it gives you the benefit of looking at things in a slightly different way. In the long run, it will not make any difference which way round you do things.

2.1 SEQUENCES

Intuitively, a sequence (in the mathematical sense, which is to say an *infinite* sequence) is a list of objects that goes on forever. The formal definition is in terms of functions.

2.1.1 Definition

A *sequence* is a function whose domain is the set of positive integers. If f is such a function, we can generate all its values one by one,

$$f(1),\ f(2),\ f(3),\ \text{etc.}$$

and this gives us a sequence in the everyday sense of the word. The values appear one after the other, 'in sequence'.

A common way to denote a sequence uses dots, like this:

$$f(1),\ f(2),\ f(3),\ \ldots$$

The three dots indicate that the terms continue forever. For example,

$$1, 4, 9, 16, 25,\ \ldots$$

denotes the sequence of all squares of positive integers.

If we use the symbol s_n to denote the value $f(n)$ of the function f, the sequence of values is denoted by

$$s_1, s_2, s_3,\ \ldots, s_n,\ \ldots$$

This is often abbreviated using the notation $\{s_n\}$.

For example, $\{2n-1\}$ denotes the sequence of odd positive integers

$$1,\ 3,\ 5,\ 7,\ \ldots$$

The expression $2n-1$ is called the *general term* of this sequence.

Similarly, $\left\{\dfrac{1}{n^2}\right\}$ denotes the sequence of reciprocals of the squares of all positive integers:

$$\frac{1}{1},\ \frac{1}{4},\ \frac{1}{9},\ \frac{1}{16},\ \ldots$$

The general term is $\frac{1}{n^2}$.

2.1.2 Limits of Sequences

For the sequence

$$1\tfrac{1}{1},\ 1\tfrac{1}{2},\ 1\tfrac{1}{3},\ \ldots,\ 1\tfrac{1}{n},\ \ldots$$

the terms in the sequence get progressively closer to the value 1. For any given measure of closeness, say 1/1000, then from some point on (namely from the 1001st term on), every term in the sequence is within that distance of 1. We say that 1 is the *limit* of the sequence.

Formally, a sequence $\{s_n\}$ has *limit* L if, for any given positive real number ϵ (epsilon), however small, from some point on every term in the sequence is within a distance ϵ of L.

Expressing this definition in algebraic form, the sequence $\{s_n\}$ has limit L if, for any positive real number ϵ, however small, there is an integer N such that $|s_n - L| < \epsilon$, for all $n \geq N$.

In the above definition, the N will, in general, depend on the ϵ ; the smaller the ϵ, the larger N will have to be.

There are two common notations to indicate that a sequence $\{s_n\}$ has limit L:

$$\lim_{n\to\infty} s_n = L$$

and

$$s_n \to L \text{ as } n \to \infty$$

You read the first of these as "the limit of s-sub-n as n tends to infinity is L." You read the second as "s-sub-n tends to L as n tends to infinity."

For example,

$$\lim_{n\to\infty}\left[1+\frac{1}{n}\right] = 1$$

and

$$(0.5)^n \to 0 \text{ as } n \to \infty$$

Many sequences do not have a limit. For example, neither of the following sequences has a limit:

$$2,\ 4,\ 6,\ \ldots,\ 2n,\ \ldots$$

$$1,\ -1,\ 1,\ -1,\ 1,\ \ldots,\ (-1)^{n+1},\ \ldots$$

A sequence that has a limit is said to *converge* to that limit, and is called a *convergent* sequence.

A sequence that does not have a limit is said to *diverge* and to be a *divergent* sequence.

The limit of a convergent sequence does not have to occur in the sequence. For example, the limit 0 of the sequence $\{1/n\}$ is not a term in the sequence.

2.1.3 Infinite Limits

If the members of a sequence $\{s_n\}$ become arbitrarily large as n gets bigger, we say that $\{s_n\}$ *approaches infinity* as n goes to infinity. This is written symbolically as:

$$s_n \to \infty \text{ as } n \to \infty$$

or

$$\lim_{n\to\infty} s_n = \infty$$

Formally, this means that for any given positive number M, however large, there is a number N such that $s_n > M$ whenever $n > N$.

For example, $\sqrt{n} \to \infty$ as $n \to \infty$. Given any positive number M, if we take N at least as big as M^2, then $\sqrt{n} > M$ whenever $n > N$.

If the members of a sequence $\{s_n\}$ get arbitrarily large negatively as n becomes bigger, we say that the sequence *approaches negative infinity* as n approaches infinity, and write

$$s_n \to -\infty \quad \text{as} \quad n \to \infty$$

or

$$\lim_{n\to\infty} s_n = -\infty$$

Formally, for any given positive number M, there is a number N such that $s_n < -M$ for all $n > N$.

For example,

$$\lim_{n\to\infty} -\ln n = -\infty$$

To show this, suppose we are given an $M > 0$. Take $N = e^M$. Then for any $n > N$, $\ln n > M$, so $-\ln n < -M$, as the definition requires.

2.1.4 Examples

✎ ***Example 1***

Use the definition of limit to prove that

$$\lim_{n\to\infty} \frac{2n}{n+1} = 2$$

Solution

Let $\epsilon > 0$ be given. We need to find an N such that

$$\left|\frac{2n}{n+1} - 2\right| < \epsilon$$

whenever $n \geq N$.

Rewrite the above inequality as

$$\left|\frac{2n - 2(n+1)}{n+1}\right| < \epsilon$$

that is,

$$\left|\frac{-2}{n+1}\right| < \epsilon$$

which simplifies to

$$\frac{2}{n+1} < \epsilon$$

Let N be the smallest integer such that $N > \frac{2}{\epsilon} - 1$. Then, whenever $n \geq N$, we have $n > \frac{2}{\epsilon} - 1$, so $n + 1 > \frac{2}{\epsilon}$, so $\frac{2}{n+1} < \epsilon$, and hence (by the above)

$$\left|\frac{2n}{n+1} - 2\right| < \epsilon$$

✎ ***Example 2***

Find sequences $\{a_n\}$, $\{b_n\}$ such that

$$\lim_{n\to\infty} a_n = \infty \qquad \lim_{n\to\infty} b_n = \infty \qquad \lim_{n\to\infty} (a_n - b_n) = 1$$

Solution

Let $a_n = n + 1$, $b_n = n$. Then $a_n - b_n = 1$, for all n, so the result is immediate.

2.1.5 Problems

1. Write down the first four terms of each sequence:

   (a) $\left\{\frac{1}{n(n+1)}\right\}$ **(b)** $\left\{\frac{2^n}{n^2}\right\}$ **(c)** $\left\{(-1)^n \frac{n}{n+1}\right\}$

2. Write down a general term for each sequence:

   (a) $\frac{2}{3}, \frac{4}{5}, \frac{6}{7}, \frac{8}{9}, \frac{10}{11}, \ldots$

(b) $\frac{1}{3}, -\frac{1}{5}, \frac{1}{7}, -\frac{1}{9}, \frac{1}{11}, -\frac{1}{13}, \ldots$

(c) $\frac{1.2}{3.4}, \frac{2.3}{4.5}, \frac{3.4}{5.6}, \frac{4.5}{6.7}, \frac{5.6}{7.8}, \ldots$

3. Determine the limit, if any, of each sequence in Problems 1 and 2 on page 31 and above.

4. Use the definition of limit to show that

$$\lim_{n\to\infty} \frac{1}{\sqrt{n}} = 0$$

5. Try to find sequences $\{a_n\}$, $\{b_n\}$ such that

$$\lim_{n\to\infty} a_n, \quad \lim_{n\to\infty} b_n$$

exist (and are finite) but

$$\lim_{n\to\infty} (a_n + b_n) = \infty$$

Either find such sequences or show that it is impossible.

6. Try to find sequences $\{a_n\}$, $\{b_n\}$ of nonzero numbers such that

$$\lim_{n\to\infty} a_n, \quad \lim_{n\to\infty} b_n$$

exist (and are finite) but

$$\lim_{n\to\infty} \frac{a_n}{b_n} = \infty$$

Either find such sequences or show that it is impossible.

2.2 LIMITS OF FUNCTIONS

Though the concept of a limit of a function (as the argument approaches a particular real number) is different from the concept of the limit of a sequence (as n approaches infinity), the formal definitions of the two concepts are similar. In fact, there are close connections between the two notions of limit.

2.2.1 Definition

Let f be a function from real numbers to real numbers. For a given real number a, we say that a real number L is the *limit* of $f(x)$ as x tends to a if $f(x)$ becomes arbitrarily close to L as x gets closer to a.

Formally, for any given positive number ϵ (epsilon), however small, there is a positive number δ (delta) such that

$$|f(x) - L| < \epsilon, \quad \text{for all } x \text{ such that } \quad 0 < |x - a| < \delta$$

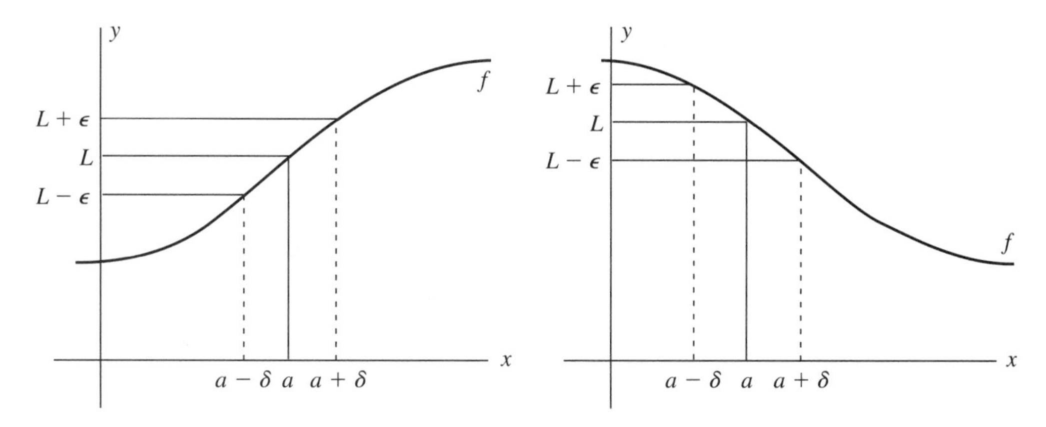

FIGURE 2.1 *Definition of a limit*

Eliminating the absolute value function, this can be rewritten as

$$L - \epsilon < f(x) < L + \epsilon, \quad \text{for all } x \text{ such that} \quad a - \delta < x < a + \delta, \ x \neq a$$

See Figure 2.1. (The figure is not comprehensive; it shows only two cases, where the function is strictly increasing and where it is strictly decreasing.)

In the above definition, the order in which things happen is important. To show that L is the limit of $f(x)$ as x approaches a, you have to show that, for any given $\epsilon > 0$, $f(x)$ is within ϵ of L for all x sufficiently close to a. First comes ϵ. Then you pick a δ that tells you how close x has to be to a for $f(x)$ to be within ϵ of L. The δ depends on ϵ. The smaller ϵ, the smaller δ may have to be.

The usual notation for limits is

$$\lim_{x \to a} f(x) = L$$

You may also see this notation:

$$f(x) \to L \text{ as } x \to a$$

In the formal definition, $f(a)$ does not have to be equal to L. In fact, f does not even have to be defined at $x = a$. Given an ϵ, when the δ has been chosen, we have to look at only those arguments x for which

$$0 < |x - a| < \delta$$

The "$0 <$" condition explicitly excludes $x = a$. This is important. The key to the differential calculus is being able to evaluate limits at points where the function is undefined.

For some functions f, the values $f(x)$ approach a fixed limit value as x approaches ∞ or $-\infty$. This gives rise to the limits

$$\lim_{x\to\infty} f(x) \text{ and } \lim_{x\to-\infty} f(x)$$

Here are the definitions of these limits:

- $\lim_{x\to\infty} f(x) = L$ if, for any $\epsilon > 0$, there is a real number M such that
$$|f(x) - L| < \epsilon \text{ whenever } x > M.$$
- $\lim_{x\to-\infty} f(x) = L$ if, for any $\epsilon > 0$, there is a real number M such that
$$|f(x) - L| < \epsilon \text{ whenever } x < M.$$

2.2.2 Infinite Limits

For some functions f, the values $f(x)$ get arbitrarily large as x approaches a fixed number a. In such a case, we say that the function approaches ∞ as x approaches a, and we write

$$\lim_{x\to a} f(x) = \infty$$

For example, the function $1/|x|$ approaches ∞ as x approaches 0.

Formally, $\lim_{x\to a} f(x) = \infty$ if, given any positive number M, however large, there is a positive number δ such that

$$f(x) > M \text{ whenever } 0 < |x - a| < \delta.$$

Similarly, if the values $f(x)$ of a function f become arbitrarily large negatively as x approaches a fixed number a, we say that $f(x)$ approaches $-\infty$ as x approaches a, and write

$$\lim_{x\to a} f(x) = -\infty$$

Formally, $\lim_{x\to a} f(x) = -\infty$ if, given any negative number M, however large, there is a positive number δ such that

$$f(x) < M, \text{ whenever } 0 < |x - a| < \delta.$$

2.2.3 Directional Limits

If $f(x)$ gets arbitrarily close to L as x approaches a from the left (i.e. moving right), we say that $f(x)$ approaches L as x approaches *a from the left*, and write

$$\lim_{x\to a-} f(x) = L$$

Here is the formal definition: Given any positive number ϵ, however small, there is a positive number δ such that $|f(x) - L| < \epsilon$ whenever $x < a$ and $0 < |x - a| < \delta$.

If $f(x)$ gets arbitrarily close to L as x approaches a from the right (i.e. moving left), we say that $f(x)$ approaches L as x approaches a *from the right*, and write

$$\lim_{x \to a+} f(x) = L$$

Here is the formal definition: Given any positive number ϵ, however small, there is a positive number δ such that $|f(x) - L| < \epsilon$ whenever $x > a$ and $0 < |x - a| < \delta$.

It can happen that $\lim_{x \to a+} f(x)$ and $\lim_{x \to a-} f(x)$ both exist but are not equal. In this case, $\lim_{x \to a} f(x)$ does not exist.

2.2.4 Theorems on Limits

Theorem. Suppose that

$$\lim_{x \to a} f(x) = L, \ \lim_{x \to a} g(x) = M$$

where a is either a real number or $\pm\infty$. Then:

1. $\lim\limits_{x \to a} kf(x) = kL$ (k any constant)
2. $\lim\limits_{x \to a}[f(x) \pm g(x)] = L \pm M$
3. $\lim\limits_{x \to a}[f(x)g(x)] = LM$
4. $\lim\limits_{x \to a}[f(x)/g(x)] = L/M \qquad (M \neq 0)$
5. $\lim\limits_{x \to a}[f(x)]^n = L^n \qquad (n = 1, 2, \ldots)$

Proofs

1. If $k = 0$, the result is trivial, so suppose $k \neq 0$.
   Let $\epsilon > 0$ be given. Then $\dfrac{\epsilon}{|k|} > 0$. (You'll see why we consider $\dfrac{\epsilon}{|k|}$ when we reach the end of the argument.) So, as $\lim\limits_{x \to a} f(x) = L$, we have

$$|f(x) - L| < \frac{\epsilon}{|k|}, \text{ whenever } 0 < |x - a| < \delta$$

   Then, for $0 < |x - a| < \delta$,

$$|kf(x) - kL| = |k|\,|f(x) - L| < |k|\frac{\epsilon}{|k|} = \epsilon$$

   This shows that $\lim\limits_{x \to a} kf(x) = kL$, as required.

2. We are given $\epsilon > 0$. Choose $\delta_1 > 0$ so that

$$|f(x) - L| < \frac{\epsilon}{2} \text{ whenever } 0 < |x - a| < \delta_1$$

[Using $\epsilon/2$ as the basis for choosing δ_1, rather than ϵ, may seem mysterious. It is arrived at by working backward from the desired inequality. When you reach the end of the argument, you will see why $\epsilon/2$ was chosen.]

Likewise, choose $\delta_2 > 0$ so that

$$|g(x) - M| < \frac{\epsilon}{2}, \text{ whenever } 0 < |x - a| < \delta_2$$

Let $\delta = \min(\delta_1, \delta_2)$. [Why do you think we take the minimum of the two δ s?] Then, whenever $0 < |x - a| < \delta$, we have:

$$|f(x) - L| < \frac{\epsilon}{2} \text{ and } |g(x) - M| < \frac{\epsilon}{2}$$

So, using the triangle inequality, whenever $0 < |x - a| < \delta$,

$$\begin{aligned} |(f(x) + g(x)) - (L + M)| &= |(f(x) - L) + (g(x) - M)| \\ &\leq |f(x) - L| + |g(x) - M| \\ &< \frac{\epsilon}{2} + \frac{\epsilon}{2} = \epsilon \end{aligned}$$

as required. (Now can you see why $\epsilon/2$ was used, and why we took δ to be the minimum of δ_1 and δ_2?)

3. Here is the main idea. We are given $\epsilon > 0$, and we want to choose a $\delta > 0$ so that

$$|f(x)g(x) - LM| < \epsilon, \text{ whenever } 0 < |x - a| < \delta$$

The key to choosing δ is to rewrite $(f(x)g(x) - LM)$ as

$$(f(x) - L)(g(x) - M) + M(f(x) - L) + L(g(x) - M)$$

Then, using the triangle inequality,

$$|f(x)g(x)-LM| \leq |f(x)-L|\,|g(x)-M|+|M|\,|f(x)-L|+|L|\,|g(x)-M|$$

The idea now is to choose δ so that each of the three terms on the right is less than $\epsilon/3$ whenever $0 < |x - a| < \delta$.

Choose $\delta_1 > 0$ so that

$$|f(x) - L| < \sqrt{\frac{\epsilon}{3}}, \text{ whenever } 0 < |x - a| < \delta_1$$

Choose $\delta_2 > 0$ so that

$$|g(x) - M| < \sqrt{\frac{\epsilon}{3}}, \text{ whenever } 0 < |x - a| < \delta_2$$

Choose $\delta_3 > 0$ so that

$$|f(x) - L| < \frac{\epsilon}{3(|M|+1)}, \quad \text{whenever } 0 < |x-a| < \delta_3$$

Choose $\delta_4 > 0$ so that

$$|g(x) - M| < \frac{\epsilon}{3(|L|+1)}, \quad \text{whenever } 0 < |x-a| < \delta_4$$

Let $\delta = \min(\delta_1, \delta_2, \delta_3, \delta_4)$. Then, whenever $0 < |x-a| < \delta$:

$$|f(x) - L|\,|g(x) - M| < \sqrt{\frac{\epsilon}{3}}\sqrt{\frac{\epsilon}{3}} = \frac{\epsilon}{3}$$

$$|M|\,|f(x) - L| < \frac{|M|\epsilon}{3(|M|+1)} < \frac{\epsilon}{3}$$

$$|L|\,|g(x) - M| < \frac{|L|\epsilon}{3(|L|+1)} < \frac{\epsilon}{3}$$

as required.

4. Since $\dfrac{f(x)}{g(x)} = f(x)\dfrac{1}{g(x)}$, the desired result follows from the rule for the limit of products if we can show that

$$\lim_{x \to a} \frac{1}{g(x)} = \frac{1}{M}$$

since we would then have:

$$\begin{aligned}\lim_{x \to a} \frac{f(x)}{g(x)} &= \lim_{x \to a}\left[f(x)\frac{1}{g(x)}\right] \\ &= \left[\lim_{x \to a} f(x)\right]\left[\lim_{x \to a}\frac{1}{g(x)}\right] \\ &= [L]\left[\frac{1}{M}\right] = \frac{L}{M}\end{aligned}$$

So, given $\epsilon > 0$, we have to find a $\delta > 0$ so that

$$\left|\frac{1}{g(x)} - \frac{1}{M}\right| < \epsilon, \quad \text{whenever } 0 < |x-a| < \delta$$

First, choose $\delta_1 > 0$ so that

$$|g(x) - M| < \frac{|M|}{2}, \quad \text{whenever } 0 < |x-a| < \delta_1$$

Then, whenever $0 < |x-a| < \delta_1$, $g(x)$ is within a distance $\dfrac{|M|}{2}$ of M, so we have $|g(x)| > \dfrac{|M|}{2}$, and in particular, $g(x) \neq 0$.

Now choose $\delta > 0$ so that $\delta \leq \delta_1$ and

$$|g(x) - M| < \frac{\epsilon|M|^2}{2}, \quad \text{whenever } 0 < |x-a| < \delta$$

Then, whenever $0 < |x - a| < \delta$, we have

$$\begin{aligned}\left|\frac{1}{g(x)} - \frac{1}{M}\right| &= \left|\frac{M - g(x)}{g(x)M}\right| \\ &= \frac{|M - g(x)|}{|g(x)M|} \\ &= \frac{|M - g(x)|}{|g(x)|\,|M|} \\ &= |M - g(x)|\,\frac{1}{|g(x)|}\,\frac{1}{|M|} \\ &\leq \frac{\epsilon|M|^2}{2}\,\frac{2}{|M|}\,\frac{1}{|M|} = \epsilon\end{aligned}$$

as required.

5. This result follows by repeated application of result 3. □

2.2.5 Computing Limits

Using the above theorem, you can compute limits of many functions. In particular, if $f(x)$ is a polynomial, then

$$\lim_{x \to a} f(x) = f(a)$$

For example,

$$\lim_{x \to 2}[x^3 + 7x - 1] = (2)^3 + 7(2) - 1 = 21$$

To compute a limit of a more complicated expression (when it is possible), you might have to manipulate the expression algebraically. For example, to compute

$$\lim_{x \to \infty} \frac{x^2 + 1}{3x^2 - 2x}$$

first divide both numerator and denominator in the expression by x^2, transforming the problem to

$$\lim_{x \to \infty} \frac{1 + (1/x^2)}{3 - (2/x)}$$

As $x \to \infty$, both $1/x^2$ and $2/x$ tend to 0, so

$$\lim_{x \to \infty} \frac{1 + (1/x^2)}{3 - (2/x)} = \frac{1 + 0}{3 - 0} = \frac{1}{3}$$

Notice that it is possible for a function to have a limit at a point where the function itself is not defined. For example, consider the function

$$f(x) = \frac{x - 1}{x^2 - 1}$$

This function is not defined for $x = 1$. (The expression evaluates as the indeterminate $\frac{0}{0}$.) However, by writing $x^2 - 1$ as $(x+1)(x-1)$ we have

$$f(x) = \frac{x-1}{(x+1)(x-1)}$$

Provided $x - 1 \neq 0$ (i.e., provided $x \neq 1$), we can cancel $x - 1$ from numerator and denominator to obtain

$$f(x) = \frac{1}{x+1}, \qquad \text{for } x \neq 1.$$

Since

$$\lim_{x \to 1} \frac{1}{x+1} = \frac{1}{2}$$

we have (noting that evaluation of the limit does not require computation of $f(1)$)

$$\lim_{x \to 1} f(x) = \frac{1}{2}$$

2.2.6 Two Trig Limits

All applications of the calculus to trig functions depend on the two fundamental limit results we'll prove in this section.

Theorem. If the angle x is measured in radians, then

1. $\lim_{x \to 0} \frac{\sin x}{x} = 1$
2. $\lim_{x \to 0} \frac{\cos x - 1}{x} = 0$

Proof

1. Since

$$\frac{\sin(-x)}{-x} = \frac{-\sin x}{-x} = \frac{\sin x}{x}$$

we need only consider the case $\lim_{x \to 0+} \frac{\sin x}{x}$

Referring to Figure 2.2, let AB be a small arc of the unit circle centered at O of angle x radians. Let BC be the perpendicular from B to OA. Let CD be the arc of the circle drawn with center O from C to meet OB at D.

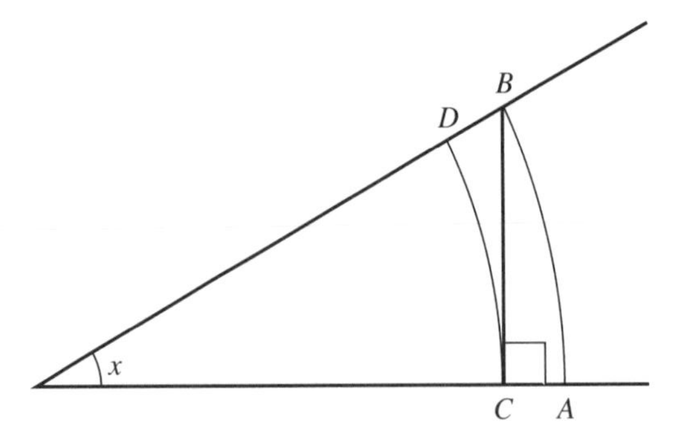

FIGURE 2.2 *A fundamental trigonometric limit*

Then, since $OB = 1$, we have $OC = \cos x$ and $BC = \sin x$. The area of the circle of radius OA is $\pi[OA]^2 = \pi$, so the area of the sector AOB is

$$\frac{x}{2\pi}\pi = \frac{x}{2}$$

The circle of radius OC has area $\pi[OC]^2 = \pi\cos^2 x$, so the area of the sector COD is

$$\frac{x}{2\pi}\pi\cos^2 x = \frac{x}{2}\cos^2 x$$

The area of the triangle COB is

$$\frac{1}{2}[OC][BC] = \frac{1}{2}\cos x \sin x$$

Comparing areas,

$$\text{sector } COD \leq \text{ triangle } COB \leq \text{ sector } AOB$$

Thus,

$$\frac{x}{2}\cos^2 x \leq \frac{1}{2}\cos x \sin x \leq \frac{x}{2}.$$

Dividing through by $\frac{x}{2}\cos x$,

$$\cos x \leq \frac{\sin x}{x} \leq \frac{1}{\cos x}$$

As $x \to 0+$, $\cos x \to 1$ and $1/\cos x \to 1$, so taking limits,

$$\lim_{x\to 0+}\frac{\sin x}{x} = 1$$

That completes the proof of part 1.

2. We deduce the second limit result from the first. The key idea is to multiply both numerator and denominator by the expression $\cos x + 1$.

$$\begin{aligned}
\lim_{x\to 0}\left[\frac{\cos x - 1}{x}\right] &= \lim_{x\to 0}\left[\frac{\cos x - 1}{x}\,\frac{\cos x + 1}{\cos x + 1}\right] \\
&= \lim_{x\to 0}\left[\frac{\cos^2 x - 1}{x(\cos x + 1)}\right] \\
&= \lim_{x\to 0}\left[\frac{-\sin^2 x}{x(\cos x + 1)}\right] \qquad (\text{since } \sin^2 x + \cos^2 x = 1) \\
&= -\lim_{x\to 0}\left[\frac{\sin x}{x}\,\frac{\sin x}{\cos x + 1}\right] \\
&= -\left[\lim_{x\to 0}\frac{\sin x}{x}\right]\left[\lim_{x\to 0}\frac{\sin x}{\cos x + 1}\right] \\
&= -(1)\left(\frac{0}{2}\right) = 0
\end{aligned}$$

That completes the proof of part 2. □

2.2.7 Examples

✎ ***Example 1***

Investigate

$$\lim_{x\to 2}\frac{x-2}{x^2-4}$$

Solution

The function is undefined for $x = 2$. For $x \neq 2$, we have

$$\frac{x-2}{x^2-4} = \frac{x-2}{(x+2)(x-2)} = \frac{1}{x+2}$$

so

$$\lim_{x\to 2}\frac{x-2}{x^2-4} = \lim_{x\to 2}\frac{1}{x+2} = \frac{1}{4}$$

✎ ***Example 2***

Investigate

$$\lim_{x\to 0}\frac{1}{1+2^{1/x}}$$

Solution

As $x \to 0+$, $1/x \to \infty$, so $2^{1/x} \to \infty$, and hence $\displaystyle\lim_{x\to 0+}\frac{1}{1+2^{1/x}} = 0.$

As $x \to 0-$, $1/x \to -\infty$, so $2^{1/x} \to 0$, and hence $\displaystyle\lim_{x\to 0-}\frac{1}{1+2^{1/x}} = 1.$

Since the left- and right-hand limits are unequal, the given limit cannot exist.

✎ Example 3

Using the definition of limit, prove that

$$\lim_{x \to 2}(x^2 + 3) = 7$$

Solution

Let $\epsilon > 0$ be given. We need to find a $\delta > 0$ such that

$$|(x^2 + 3) - 7| < \epsilon, \text{ whenever } 0 < |x - 2| < \delta.$$

i.e., so that

$$|x^2 - 4| < \epsilon, \text{ whenever } 0 < |x - 2| < \delta.$$

Let δ be the minimum of $\sqrt{\dfrac{\epsilon}{2}}, \dfrac{\epsilon}{8}$. (The reason for this rather strange-looking choice of δ will become clear when we reach the end of the proof.) Then, for $0 < |x - 2| < \delta$, we have

$$\begin{aligned}
|x^2 - 4| &= |(x-2)^2 + 4x - 8| \\
&= |(x-2)^2 + 4(x-2)| \\
&\leq |(x-2)^2| + |4(x-2)| \\
&= |x-2|^2 + 4|x-2| \\
&< \left(\sqrt{\frac{\epsilon}{2}}\right)^2 + 4\left(\frac{\epsilon}{8}\right) \\
&\qquad\qquad \text{(because of our choice of } \delta\text{)} \\
&= \frac{\epsilon}{2} + \frac{\epsilon}{2} \\
&= \epsilon
\end{aligned}$$

✎ Example 4

Using the definition of limit, prove that

$$\lim_{x \to \infty} \frac{x}{x - 1} = 1$$

Solution

Let $\epsilon > 0$ be given. We need to find a number $M > 0$ such that

$$\left|\frac{x}{x-1} - 1\right| < \epsilon, \text{ whenever } x > M,$$

i.e., such that

$$\left|\frac{x-(x-1)}{x-1}\right| < \epsilon, \text{ whenever } x > M,$$

which simplifies to

$$\left|\frac{1}{x-1}\right| < \epsilon, \text{ whenever } x > M.$$

Let $M = \dfrac{1}{\epsilon} + 1$. Then, whenever $x > M$, we have

$$x > \frac{1}{\epsilon} + 1$$

$$x - 1 > \frac{1}{\epsilon}$$

$$0 < \frac{1}{x-1} < \epsilon$$

as required.

2.2.8 Problems

1. Evaluate the following limits:

   (a) $\lim_{x \to 3}(x^2 - 9x + 1)$

   (b) $\lim_{x \to 1}(x^3 - 3x^2 + 2x + 1)$

   (c) $\lim_{x \to 3} \dfrac{x-3}{x^2-9}$

   (d) $\lim_{x \to 1+} \dfrac{x-1}{\sqrt{x^2-1}}$

   (e) $\lim_{x \to 2+} \dfrac{\sqrt{x-2}}{x^2-4}$

2. Investigate

$$\lim_{x \to \infty} \frac{3 + 2^{1/x}}{x^2}$$

3. Using the definition of a limit, show that

$$\lim_{x \to 2}(x^2 - 3x + 1) = -1$$

4. Evaluate the following limits:

(a) $\lim_{x \to 0} \frac{\sin 2x}{x}$

(b) $\lim_{x \to 0} \frac{\sin 3x}{\sin 5x}$

(c) $\lim_{x \to 0} \frac{\sin^3 5x}{\sin^2 3x}$

(Hint: For each limit, find a way to use the theorem in Section 2.2.6.)

5. Using a graphing calculator or a computer algebra system:

(a) Draw a graph of the function $\frac{\sin x}{x}$ for $-2\pi \le x \le 2\pi$, and note the behavior of the function near $x = 0$.

(b) With your calculator in radian mode, compute values of $\frac{\sin x}{x}$ for $x = 0.1,\ 0.01,\ 0.001$, etc.

(c) Do your results confirm the theorem $\lim_{x \to 0} \frac{\sin x}{x} = 1$?

6. Using a graphing calculator or a computer algebra system:

(a) Draw a graph of the function x^x for $0 < x \le 1$. By zooming in on the graph around $x = 0$, examine the behavior of the function near $x = 0$.

(b) Compute values of x^x for $x = 0.1,\ 0.01,\ 0.001$, etc.

(c) What can you say about $\lim_{x \to 0+} x^x$?

7. Using a graphing calculator or a computer algebra system:

(a) Draw a graph of the function $(1+x)^{1/x}$ for $-1 \le x \le 1$. By zooming in on the graph around $x = 0$, examine the behavior of the function near $x = 0$.

(b) Compute values of $(1+x)^{1/x}$ for $x = 0.1,\ 0.01,\ 0.001$, etc.

(c) What can you say about $\lim_{x \to 0+} (1+x)^{1/x}$?

2.3 CONTINUITY

The key operation of the differential calculus is called differentiation. It's a process that operates on functions to produce new functions. The functions on which differentiation works have to be *continuous*. Intuitively (but not completely accurately), a function is continuous if its graph has no breaks. The formal definition, which we give next, depends on the theory of limits.

2.3.1 Definition

A function $f(x)$ is *continuous* at a point a if:

1. $f(a)$ is defined;
2. $\lim_{x \to a} f(x)$ exists;
3. $\lim_{x \to a} f(x) = f(a)$.

If one or more of the above conditions fails, we say $f(x)$ is *discontinuous* at a.

If $f(x)$ is continuous at every point a in its domain, we say, simply, $f(x)$ is *continuous.*

Here is the formal ϵ, δ definition of continuity: A function $f(x)$ is *continuous* at a point a if, given any positive real number ϵ, however small, there is a positive real δ such that:

$$|f(x) - f(a)| < \epsilon, \quad \text{whenever } |x - a| < \delta$$

2.3.2 Intermediate Value Theorem

The Intermediate Value Theorem expresses a fundamental property of continuous functions. Intuitively, it says that, for a continuous function, "You can't get from here to there without passing through all points in between."

Intermediate Value Theorem. If $f(x)$ is a continuous function defined on the closed interval $[a, b]$, and if d is any real number between $f(a)$ and $f(b)$, then there is a real number c between a and b such that $f(c) = d$. (See Figure 2.3 on page 46.) ■

2.3.3 Theorems on Continuity

The following properties of continuity are used frequently in calculus.

Theorem. Suppose $f(x)$ and $g(x)$ are continuous at a and that k is a constant. Then each of the following functions is continuous at a:

- $kf(x)$
- $f(x) + g(x)$
- $f(x) - g(x)$
- $f(x)g(x)$

If, in addition, $g(x) \neq 0$, then

- $f(x)/g(x)$

is also continuous at a. ■

These results all follow immediately from the corresponding theorems on limits in Section 2.2.4 on page 35.

2.3.4 Example

✎ ***Example 1***

Suppose that $f(x)$ is defined on an interval $(a-l, a+r)$ and is continuous at a. Show that if $f(a) > 0$, then there are real numbers λ, $\delta > 0$ such that

$$f(x) \geq \lambda, \quad \text{for all } x \in (a-\delta, a+\delta).$$

Solution

Let $f(a) = b$. Set $\epsilon = \dfrac{b}{2}$. Because $f(x)$ is continuous at a, there is a $\delta > 0$ such that

$$|f(x) - b| < \epsilon, \quad \text{whenever } |x - a| < \delta.$$

Thus, for any $x \in (a-\delta, a+\delta)$,

$$-\frac{b}{2} < f(x) - b < \frac{b}{2}$$

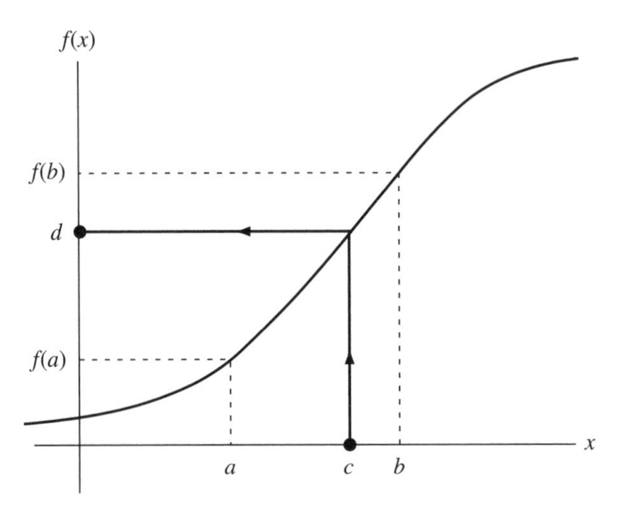

FIGURE 2.3 *The Intermediate Value Theorem*

So,

$$f(x) > \frac{b}{2}$$

which means we are done if we let $\lambda = \frac{b}{2}$.

2.3.5 Problems

1. Find all points of discontinuity of the following functions:

   (a) $x^3 - 5x^2 + 2x + 1$ (b) $\frac{(3x-4)^2}{(x+1)^3}$

   (c) $\frac{2^x - 2^{-x}}{2^x + 2^{-x}}$ (d) $\frac{x-1}{x^2+1}$

   (e) $\frac{x-2}{\sqrt{x^2-4}}$ (f) $\frac{x}{x^2-5x+6}$

   (g) $\frac{x^2-4}{x^2-5x+6}$

2. For each function in Problem 1, examine each point of discontinuity and see if it is possible to define a value for the function at that point so that the resulting function is continuous at the point. For example, the function

   $$f(x) = \frac{x-1}{x^2-1}$$

   is undefined at $x = 1$, but $\lim_{x \to 1} f(x) = \frac{1}{2}$, therefore if we *define* $f(1) = \frac{1}{2}$, the resulting function is continous at $x = 1$.

3. Use a grapher to graph each of the functions in Problem 1 around any points of discontinuity. In cases where you were able to define a continuous function in Problem 2, does your graph represent the function you defined?

4. Use a grapher to investigate the continuity of the following functions:

   (a) $f(x) = \frac{1 - 2^{1/x}}{1 + 2^{1/x}}$

   (b) $f(x) = \frac{1}{1 + 3^{1/x}}$

   (c) $f(x) = \frac{x}{1 + 3^{1/x}}$

dx

topic 3

Differentiation I

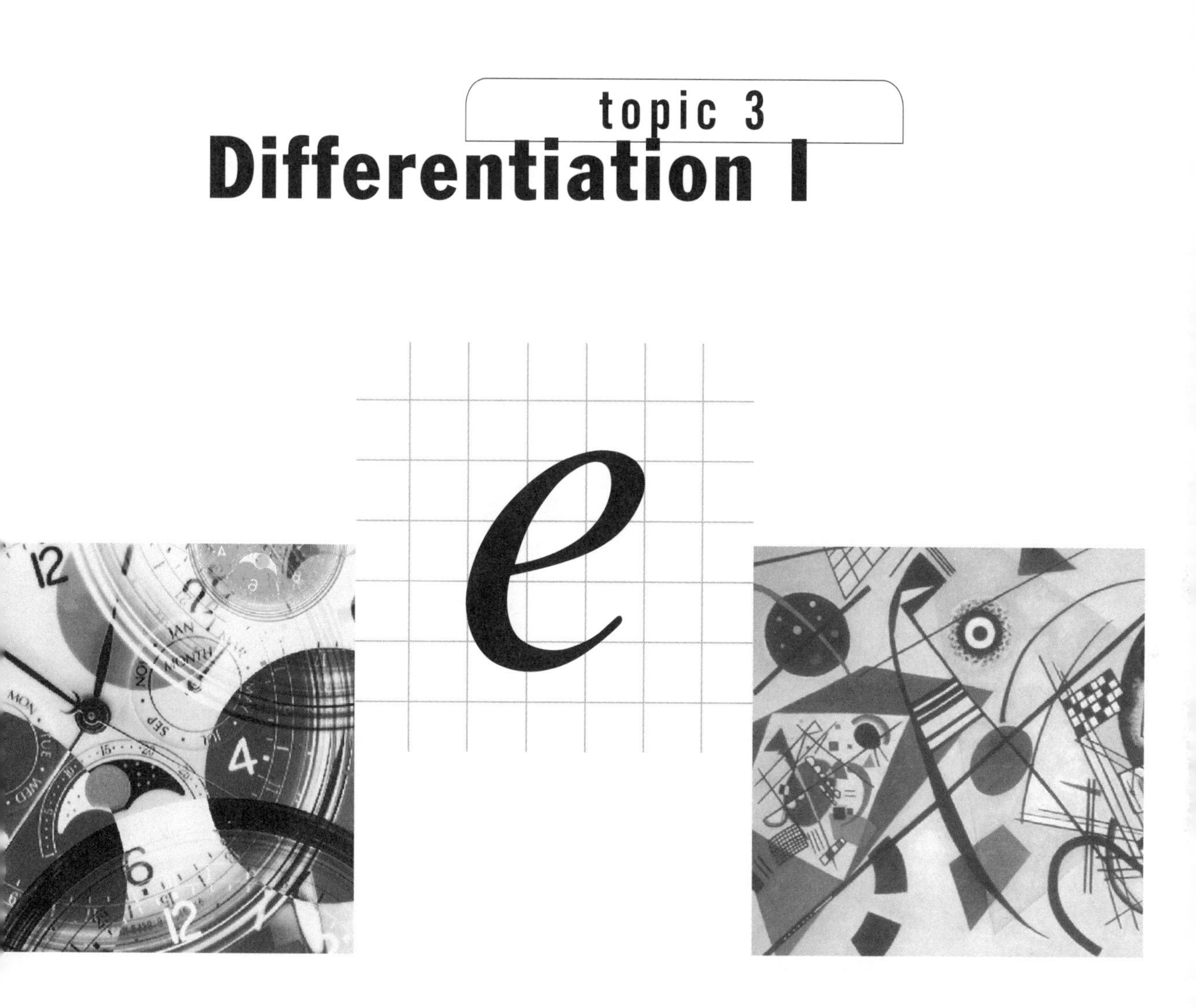

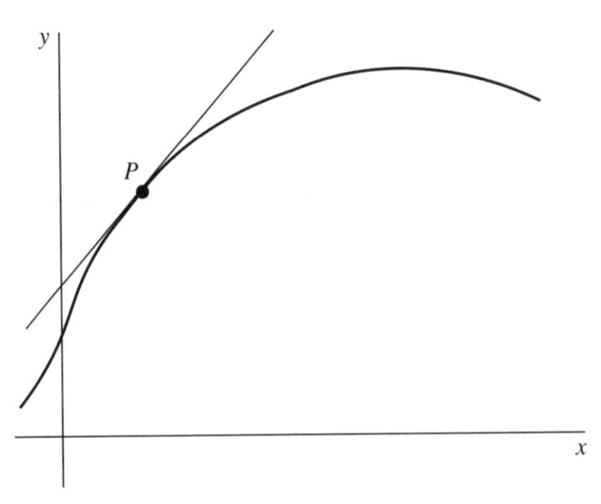

FIGURE 3.1 *Slope of a curve at a point P*

Differentiation is a method used to obtain the slope of a curve at a point P. Since the slope varies as you move along the curve, the phrase 'slope of the curve at P' is taken to mean the slope of the tangent line at P. See Figure 3.1.

3.1 THE DERIVATIVE

The idea behind differentiation is to think of trying to obtain the slope by a process of approximation. Each approximation is the slope of a straight-line chord to the curve starting at P. By taking shorter and shorter chords, the slope of the chords become closer and closer to the slope of the curve at P. The limit gives the slope. Let's take a look at the details. Figure 3.2 provides the picture.

Let P have coordinates $(a, f(a))$. The chord from P $(x = a)$ to Q $(x = a + h)$ has slope

$$\frac{f(a+h) - f(a)}{h}$$

This ratio provides an approximation to the slope of the curve at P. The smaller we make h, the better the approximation will be. Taking the limit as h approaches 0, we obtain the slope of the curve at P:

$$\lim_{h \to 0} \frac{f(a+h) - f(a)}{h}$$

provided this limit exists.

When this limit does exist, we say that the function f is *differentiable* at $x = a$. The limit is then called the *derivative* of the function f

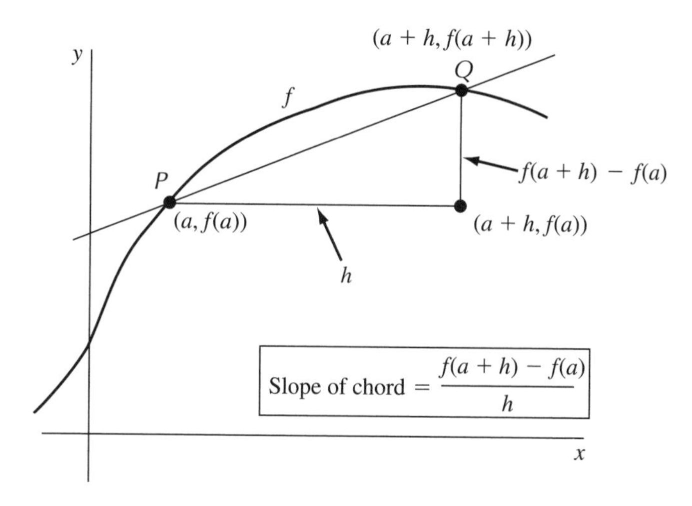

FIGURE 3.2 *Definition of the derivative*

at $x = a$. The derivative of f at $x = a$ depends on a, and is generally denoted by $f'(a)$, pronounced "f prime of a."

Notice that for each point a in the domain of f at which f is differentiable, we get a unique derivative $f'(a)$. Thus f' is itself a function. We can write it as $f'(x)$ to emphasize that it is a function. The function f' is called the *derivative function* (or simply the *derivative*) of $f(x)$.

Finding the *function* $f'(x)$ from a given function $f(x)$ is the key process of *differentiation*. Much of calculus involves learning techniques for finding derivatives of functions.

Notice that the domain of the derivative function $f'(x)$ is the set of all points x at which $f(x)$ is differentiable.

3.1.1 Differentiability

Theorem. If $f(x)$ is differentiable at $x = a$, then $f(x)$ is continuous at $x = a$.

Proof

By definition, if f is differentiable at $x = a$, the following limit exists and is equal to $f'(a)$:

$$\lim_{x \to a} \frac{f(x) - f(a)}{x - a}$$

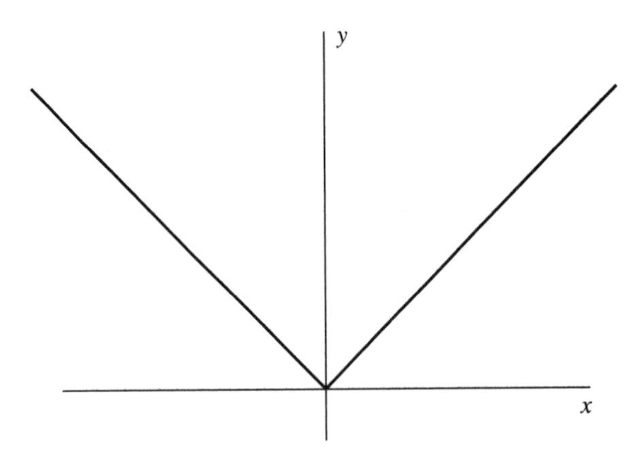

FIGURE 3.3 *The absolute value function* $y = |x|$

It follows that

$$\begin{aligned}\lim_{x\to a}[f(x) - f(a)] &= \lim_{x\to a}\left[\frac{f(x)-f(a)}{x-a} \times (x-a)\right] \\ &= \lim_{x\to a}\left[\frac{f(x)-f(a)}{x-a}\right] \times \lim_{x\to a}(x-a) \\ &= f'(a) \times 0 = 0\end{aligned}$$

Since

$$\lim_{x\to a}[f(x) - f(a)] = \lim_{x\to a} f(x) - \lim_{x\to a} f(a) = \lim_{x\to a} f(x) - f(a)$$

it follows that

$$\lim_{x\to a} f(x) - f(a) = 0$$

Hence $\lim_{x\to a} f(x) = f(a)$, which implies that $f(x)$ is continuous at $x = a$. □

The converse to the above theorem is not true. For example, the absolute value function $f(x) = |x|$ (Figure 3.3) is continuous at $x = 0$ but is not differentiable at $x = 0$ (because the slope changes abruptly from -1 to the left of 0 to $+1$ to the right of 0).

3.1.2 Notation for the Derivative

For a function given in the form

$$y = \text{formula involving } x$$

the derivative is often denoted by

$$\frac{dy}{dx}$$

Thus, for a function $y = f(x)$,

$$\frac{dy}{dx} = f'(x)$$

Note that although $\frac{dy}{dx}$ may look like a fraction (and often behaves like one as well), it's *not* a fraction; it's the derivative of y with respect to x.

A variation of the $\frac{dy}{dx}$ notation for the derivative is to use

$$\frac{d}{dx}$$

to denote the *operation* of differentiation. Thus,

$$\frac{d}{dx}(f)$$

denotes the result of differentiating the function f, namely f'. This notation emphasizes that differentiation is a process (an *operation*) that operates on one function to produce another function.

Some textbooks use a slightly different notation when defining derivatives. Suppose $y = f(x)$. If Δx denotes a small increment in x (what we denoted by h above), and we let Δy denote the corresponding increment in y, then

$$\Delta y = f(x + \Delta x) - f(x)$$

Then the definition of the derivative given above can be rewritten as

$$\frac{dy}{dx} = \lim_{\Delta x \to 0} \frac{\Delta y}{\Delta x}$$

This notation is suggestive, but it can be dangerous. The problem is this. As a "small increment," Δx is assumed to be nonzero. However, even though $\Delta x \neq 0$, it might be that $\Delta y = 0$. (This is not a rare event. For example, it happens if f is a constant function!) Thus, Δy is *not* necessarily a "small increment." In other words, Δy and Δx are not the same kind of entity. Using the same notation to represent two different kinds of object can lead to errors in mathematics, and is thus best avoided. There are, for instance, mathematics textbooks that present a false proof of the chain rule (see Section 3.2) because of the confusion between Δy and Δx.

3.1.3 Differentiation of Sums and Powers

In this section we derive rules that enable us to differentiate any polynomial function.

Theorem (Differentiation of Integer Powers). For any positive integer n,

$$\frac{d}{dx}(x^n) = nx^{n-1}$$

Proof

Before we prove the general case stated, let's see how to prove the theorem for some simple cases.

Case 1. $n = 1$. We have to show that

$$\frac{d}{dx}(x^1) = 1x^0$$

That is, we have to show that if $f(x) = x$, then $f'(x) = 1$.

By the definition of the derivative,

$$\begin{aligned} f'(x) &= \lim_{h \to 0} \frac{f(x+h) - f(x)}{h} \\ &= \lim_{h \to 0} \frac{(x+h) - (x)}{h} \\ &= \lim_{h \to 0} \frac{h}{h} \\ &= \lim_{h \to 0} 1 \\ &= 1 \end{aligned}$$

Case 2. $n = 2$. We have to show that if $f(x) = x^2$, then $f'(x) = 2x^1$.

By the definition of the derivative,

$$\begin{aligned} f'(x) &= \lim_{h \to 0} \frac{f(x+h) - f(x)}{h} \\ &= \lim_{h \to 0} \frac{(x+h)^2 - (x)^2}{h} \\ &= \lim_{h \to 0} \frac{x^2 + 2xh + h^2 - x^2}{h} && \text{(Expanding } (x+h)^2 \text{ using the binomial theorem.)} \\ &= \lim_{h \to 0} \frac{2xh + h^2}{h} && \text{(Simplifying the numerator.)} \\ &= \lim_{h \to 0}(2x + h) && \text{(Cancelling } h \text{ from numerator and denominator.)} \\ &= 2x + 0 && \text{(Limit of a sum equals the sum of the limits.)} \\ &= 2x \end{aligned}$$

as required.

Case 3. $n = 3$. We have to show that if $f(x) = x^3$, then $f'(x) = 3x^2$.

By the definition of the derivative,

$$\begin{aligned} f'(x) &= \lim_{h\to 0} \frac{f(x+h)-f(x)}{h} \\ &= \lim_{h\to 0} \frac{(x+h)^3-(x)^3}{h} \\ &= \lim_{h\to 0} \frac{x^3+3x^2h+3xh^2+h^3-x^3}{h} \end{aligned}$$

(Expanding $(x+h)^3$ using the binomial theorem.)

$$\begin{aligned} &= \lim_{h\to 0} \frac{3x^2h+3xh^2+h^3}{h} \\ &= \lim_{h\to 0}(3x^2+3xh+h^2) \\ &= 3x^2+0+0 \\ &= 3x^2 \end{aligned}$$

Case 4. General case. We have to show that if $f(x) = x^n$, then $f'(x) = nx^{n-1}$. By the definition of the derivative,

$$f'(x) = \lim_{h\to 0} \frac{f(x+h)-f(x)}{h} = \lim_{h\to 0} \frac{(x+h)^n - x^n}{h}$$

The general Binomial Theorem says that there are constants $C_2^n, \ldots, C_{n-2}^n$ (often denoted by $\binom{n}{2}, \ldots, \binom{n}{n-2}$), depending on n, such that

$$\begin{aligned} (x+h)^n &= x^n + nx^{n-1}h + C_2^n x^{n-2}h^2 + C_3^n x^{n-3}h^3 + \cdots \\ &\cdots + C_{n-3}^n x^3 h^{n-3} + C_{n-2}^n x^2 h^{n-2} + nxh^{n-1} + h^n \end{aligned}$$

Hence:

$$\begin{aligned} \frac{(x+h)^n - x^n}{h} &= nx^{n-1} + C_2^n x^{n-2}h + C_3^n x^{n-3}h^2 + \cdots \\ &\cdots + C_{n-3}^n x^3 h^{n-4} + C_{n-2}^n x^2 h^{n-3} + nxh^{n-2} + h^{n-1} \end{aligned}$$

Since all terms in the above sum except the first have a factor of h, they all approach 0 as $h \to 0$. Hence

$$\lim_{h\to 0} \frac{(x+h)^n - x^n}{h} = nx^{n-1} + 0 = nx^{n-1}$$

and we are done. □

For any constant c, the function $f(x) = c$ determines a straight line of constant slope 0, so $f'(x) = 0$ for all x. In particular, for the case $f(x) = 1$, we have

$$\frac{d}{dx}(x^0) = \frac{d}{dx}(1) = 0 = 0x^{0-1}$$

(provided $x \neq 0$), so the formula established in the above theorem also holds for $n = 0$ (assuming $x \neq 0$). In fact, it holds for any real number n:

Theorem. For any real number r,

$$\frac{d}{dx}(x^r) = rx^{r-1} \qquad (\text{for } x \geq 0). \qquad \blacksquare$$

The proof uses advanced techniques. See Examples 3.4.1, Example 2.

For example,

$$\frac{d}{dx}(x^{3/4}) = \frac{3}{4}x^{-1/4}$$

Given a function $f(x)$ and a constant a, the function af is defined by

$$[af](x) = a \times f(x), \text{ for all } x$$

Theorem (Differentiation of a Constant Multiple). If the function f is differentiable and a is a constant, then the function af is differentiable, and

$$\frac{d}{dx}[af](x) = a[f'(x)]$$

Proof

$$\begin{aligned}
\lim_{h\to 0}\frac{[af](x+h)-[af](x)}{h} &= \lim_{h\to 0}\frac{a[f(x+h)]-a[f(x)]}{h} \\
&\quad \text{(By the definition of the function } [af].) \\
&= \lim_{h\to 0} a\frac{f(x+h)-f(x)}{h} \\
&\quad \text{(Taking out the common factor } a.) \\
&= a \lim_{h\to 0}\frac{f(x+h)-f(x)}{h} \\
&\quad \text{(Since } a \text{ is a constant.)} \\
&= a\, f'(x) \qquad \square
\end{aligned}$$

Theorem (Differentiation of a Sum). If the functions f and g are differentiable, then the functions $[f \pm g]$ are differentiable, and

$$\frac{d}{dx}[f(x) \pm g(x)] = f'(x) \pm g'(x)$$

Proof

Consider the case $[f+g]$ for definiteness.

$$\begin{aligned}
\lim_{h\to 0} \frac{[f+g](x+h)-[f+g](x)}{h} \\
&= \lim_{h\to 0} \frac{[f(x+h)+g(x+h)]-[f(x)+g(x)]}{h} \\
&\qquad \text{(By definition of the function } [f+g].) \\
&= \lim_{h\to 0} \frac{[f(x+h)-f(x)]+[g(x+h)-g(x)]}{h} \\
&\qquad \text{(Rearranging the numerator.)} \\
&= \lim_{h\to 0} \left[\frac{[f(x+h)-f(x)]}{h} + \frac{[g(x+h)-g(x)]}{h} \right] \\
&\qquad \text{(Separating into two quotients.)} \\
&= \lim_{h\to 0} \frac{[f(x+h)-f(x)]}{h} + \lim_{h\to 0} \frac{[g(x+h)-g(x)]}{h} \\
&\qquad \text{(Limit of a sum equals the sum of the limits.)} \\
&= f'(x)+g'(x) \qquad \square
\end{aligned}$$

Using the rules established in the above theorems, you can differentiate any polynomial function. The idea is to differentiate term by term. For example,

$$\frac{d}{dx}\left[x^5+11x^4-8x^3+x^2+5x-3\right] = 5x^4+44x^3-24x^2+2x+5$$

3.1.4 Examples

✎ ***Example 1***

Use the definition of the derivative as a limit to determine $f'(1)$ where

$$f(x) = \frac{1}{x-2}$$

Solution

$$\begin{aligned}
f'(1) &= \lim_{h\to 0} \frac{f(1+h)-f(1)}{h} \\
&= \lim_{h\to 0} \left[\frac{1}{(1+h)-2} - \frac{1}{1-2} \right] \Big/ h \\
&= \lim_{h\to 0} \left[\frac{1}{h-1} - \frac{1}{-1} \right] \Big/ h \\
&= \lim_{h\to 0} \left[\frac{1}{h-1} + 1 \right] \Big/ h
\end{aligned}$$

$$= \lim_{h \to 0} \left[\frac{1+h-1}{h-1} \right] \Big/ h$$

$$= \lim_{h \to 0} \frac{h}{h-1} \Big/ h$$

$$= \lim_{h \to 0} \frac{1}{h-1}$$

$$= -1$$

✎ *Example 2*

Find the derivative of each of the following functions:

(a) $y = 21x^{\pi}$ **(b)** $y = \dfrac{1}{\sqrt{x}}$ **(c)** $y = \dfrac{x^{3/4}}{11}$

Solution

(a) $\dfrac{dy}{dx} = 21\pi x^{\pi - 1}$

(b) Write the function as $y = x^{-1/2}$. Then

$$\frac{dy}{dx} = -\frac{1}{2} x^{-3/2} = \frac{-1}{2x^{3/2}}$$

(c) Write the function as $y = \frac{1}{11} x^{3/4}$. Then

$$\frac{dy}{dx} = \frac{1}{11} \frac{3}{4} x^{-1/4} = \frac{3}{44} x^{-1/4} = \frac{3}{44x^{1/4}}$$

✎ *Example 3*

Find $f'(x)$, where

$$f(x) = 21x^{\pi} - \frac{1}{\sqrt{x}} + \frac{x^{3/4}}{11}$$

Deduce the slope of the graph of this function at $x = 1$.

Solution

Using the answers to Example 2,

$$f'(x) = 21\pi x^{\pi - 1} + \frac{1}{2x^{3/2}} + \frac{3}{44x^{1/4}}$$

Hence, $f'(1) = 21\pi + \frac{1}{2} + \frac{3}{44} = 21\pi + \frac{25}{44} \approx 66.5416$

3.1.5 Problems

1. Find the derivatives of the following functions:

   (a) $x^5 + 22x^3 - x^2 + 15x + 9$ **(b)** $16x^{4/7}$ **(c)** $(\pi + 1)x^{2\pi}$

   (d) $x^{-11/4}$ **(e)** $\dfrac{1}{x}$ **(f)** $\dfrac{3.141}{x^5}$ **(g)** $\dfrac{1}{\sqrt{x}}$

   (h) $\log_5(9)x^{\log_2 5}$ **(i)** π^3

2. Use the definition of the derivative as a limit to determine $f'(3)$ where

$$f(x) = \frac{1}{x-2}$$

3. Use a graphing calculator or a computer algebra system to draw a graph of the function $f(x) = x^{1/3}$ for $-1 \leq x \leq 1$. On the basis of the graph, what can you say about $f'(x)$? Sketch (by hand) a graph of $f'(x)$ for $-1 \leq x \leq 1$.

4. Use a calculator to estimate $f'(1)$ and $f'(2)$ for the function

$$f(x) = x^x$$

3.2 RULES FOR DIFFERENTIATING

So far we have derived rules that allow us to differentiate any polynomial. In this section, we develop a set of extremely powerful techniques that greatly enlarge the pool of functions we can differentiate.

3.2.1 The Chain Rule

If you have two functions, one taking x to z and another taking z to y, then by combining them you can obtain a function taking x to y: If $z = g(x)$ and $y = f(z)$, then for every argument x we first calculate z using g, and then we use that value of z to calculate y using f. The resulting function that takes you from x to y is called the *composition* of f and g, denoted by $f \circ g$. Thus,

$$[f \circ g](x) = f(g(x))$$

Composition of two functions is illustrated diagramatically in Figure 3.4 on page 62.

For example, if

$$f(z) = z^{12}$$

and

$$g(x) = x^5 + 20x^3 - 5x^2 + 1$$

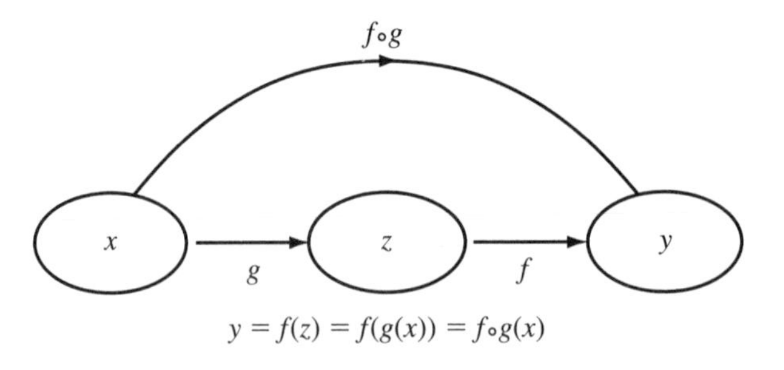

FIGURE 3.4 *Function composition*

then

$$[f \circ g](x) = (x^5 + 20x^3 - 5x^2 + 1)^{12}$$

Suppose you wanted to find the derivative of this last function. One way would be to expand the expression into a polynomial using the binomial theorem. To do so by hand would clearly involve a lot of work. The chain rule lets you find the derivative a different way. It gives you $\frac{dy}{dx}$ in terms of the two much simpler derivatives $\frac{dz}{dx}$ and $\frac{dy}{dz}$.

Theorem (The Chain Rule). If $y = f(z)$ and $z = g(x)$, and if both f and g are differentiable, then the composition $f \circ g$ is differentiable. Moreover, the derivative of

$$y = (f \circ g)(x)$$

is given by

$$\frac{dy}{dx} = \frac{dy}{dz}\frac{dz}{dx}$$

In words, the derivative of the composition is the product of the derivatives. ■

Expressed in f, g notation, the chain rule looks like this:

$$[f \circ g]'(x) = f'(z)\, g'(x)$$

For the example above, we have

$$y = z^{12}$$

so

$$\frac{dy}{dz} = 12z^{11}$$

and

$$z = x^5 + 20x^3 - 5x^2 + 1$$

so

$$\frac{dz}{dx} = 5x^4 + 60x^2 - 10x$$

Thus, by the chain rule,

$$\frac{dy}{dx} = \frac{dy}{dz}\frac{dz}{dx} = 12z^{11}(5x^4 + 60x^2 - 10x)$$

Though the above expression does give the derivative, it involves two variables, x and z, and this is unlikely to be what we want. We usually want the derivative $\frac{dy}{dx}$ to be given as a function of x. However, we can eliminate the variable z by using the original equation $z = x^5 + 20x^3 - 5x^2 + 1$ to substitute. Thus,

$$\frac{dy}{dx} = 12(x^5 + 20x^3 - 5x^2 + 1)^{11}(5x^4 + 60x^2 - 10x)$$

Proof of the Chain Rule

Suppose $y = f(z)$, $z = g(x)$, where f and g are differentiable. We prove that $f \circ g$ is differentiable, and

$$[f \circ g]'(x) = f'(z)\, g'(x)$$

or, in dy/dx notation,

$$\frac{dy}{dx} = \frac{dy}{dz}\frac{dz}{dx}$$

Some older textbooks present the following argument as (they claim) a proof of the chain rule. It is instructive to see why this argument fails.

Let Δx be a small increment in x. This produces a corresponding small increment Δz in z (by virtue of g), and that increment in turn produces a small increment Δy in y (by virtue of f). By ordinary algebra,

$$\frac{\Delta y}{\Delta x} = \frac{\Delta y}{\Delta z}\frac{\Delta z}{\Delta x}$$

Taking the limits, since $\Delta z \to 0$ as $\Delta x \to 0$,

$$\begin{aligned}\frac{dy}{dx} &= \lim_{\Delta x \to 0} \frac{\Delta y}{\Delta x} \\ &= \lim_{\Delta z \to 0} \frac{\Delta y}{\Delta z} \lim_{\Delta x \to 0} \frac{\Delta z}{\Delta x} \\ &= \frac{dy}{dz}\frac{dz}{dx}\end{aligned}$$

This is wrong. The initial step can only be made if $\Delta z \neq 0$ for all sufficiently small values of Δx. But this cannot be assumed. For example, it fails if g is a constant function! Though the conclusion is valid, the argument is not.

Let's do it right. Fix some value of x. This fixes the values $z = g(x)$ and $y = f(z)$.

To avoid the $\dfrac{\Delta y}{\Delta x}$ problem, we introduce a function $E(h)$, related to $f'(z)$, that is continuous at 0. For nonzero h, define

(1) $$E(h) = \frac{f(z+h) - f(z)}{h} - f'(z)$$

By the definition of the derivative of f,

(2) $$\lim_{h \to 0} E(h) = 0$$

So far, $E(h)$ is not defined for $h = 0$. Let's rectify this, by defining

(3) $$E(0) = 0$$

Then, for any value of h, we have

(4) $$f(z+h) - f(z) = h[E(h) + f'(z)]$$

In the case $h \neq 0$, this follows from Eq. (1) by rearranging; for $h = 0$, it is trivially true, since both sides of Eq. (4) then equal zero.

Now let Δx be a small, nonzero increment in x. The corresponding increment in z is

(5) $$\Delta z = g(x + \Delta x) - g(x)$$

(We noted earlier that Δz might be zero, so it's not an "increment" in the sense that Δx is.) Let Δy be the corresponding increment in y, i.e.,

(6) $$\Delta y = f(z + \Delta z) - f(z)$$

Using Eq. (4) with $h = \Delta z$ (and note that Eq. (4) is valid for any value of h, zero or nonzero), Eq. (6) becomes

(7) $$\Delta y = \Delta z[E(\Delta z) + f'(z)]$$

Dividing by Δx (which is not zero),

(8) $$\frac{\Delta y}{\Delta x} = \frac{\Delta z}{\Delta x}[E(\Delta z) + f'(z)]$$

Now, since g is differentiable, it is continuous, so by Eq. (5),

(9) $$\lim_{\Delta x \to 0} \Delta z = \lim_{\Delta x \to 0} [g(x + \Delta x) - g(x)] = 0$$

Thus, by Eq. (2)

(10) $$\lim_{\Delta x \to 0} E(\Delta z) = \lim_{\Delta z \to 0} E(\Delta z) = 0$$

Hence, by Eq. (8),

$$\begin{aligned}
\frac{dy}{dx} &= \lim_{\Delta x \to 0} \frac{\Delta y}{\Delta x} \\
&= \left[\lim_{\Delta x \to 0} \frac{\Delta z}{\Delta x}\right] \left[\lim_{\Delta x \to 0} [E(\Delta z) + f'(z)]\right] \\
&= \left[\frac{dz}{dx}\right] \left[0 + f'(z)\right] \qquad \text{(Using Eq. (10).)} \\
&= \frac{dz}{dx} f'(z) \\
&= \frac{dz}{dx} \frac{dz}{dx}
\end{aligned}$$

and the proof of the chain rule is complete. □

3.2.2 The Inverse Function Rule

Recall that function g is an inverse of function f if

$$y = f(x) \iff g(y) = x$$

If f and g are inverses of each other, there is a simple way to obtain g' from f' called the *Inverse Function Rule*:

Theorem (Inverse Function Rule). Let f be a differentiable function, and let g be the inverse function of f. Let a be given, and set $b = g(a)$. If $f'(b) \neq 0$, then g is differentiable at a, and

$$g'(a) = 1/f'(b)$$

In dy/dx notation, if $y = f(x)$ and $x = g(y)$, then

$$\frac{dy}{dx} = 1 \bigg/ \frac{dx}{dy}$$

Proof

By setting $h = x - a$, we can write the definition of $g'(a)$ as

$$g'(a) = \lim_{x \to a} \frac{g(x) - g(a)}{x - a}$$

provided this limit exists. For any x, let $y = f(x)$. Since $g = f^{-1}$,

$$y = f(x) \Leftrightarrow g(y) = x$$

We shall assume the (valid) result that if a function is continuous and has an inverse, then the inverse is also continuous. In this case, since f is differentiable, it is continuous, and hence g is continuous. Thus, as $x \to a$, we have $g(x) \to g(a)$, that is, $y \to b$. Moreover, since f and g both have inverses, they are both one-to-one. Thus, if $x \neq a$, then $y \neq b$, and vice versa. Hence,

$$\begin{aligned}
\lim_{x \to a} \frac{g(x) - g(a)}{x - a} &= \lim_{y \to b} \frac{y - b}{f(y) - f(b)} \\
&= \lim_{y \to b} \left[1 \bigg/ \frac{f(y) - f(b)}{y - b} \right] \\
&= 1 \bigg/ \left[\lim_{y \to b} \frac{f(y) - f(b)}{y - b} \right] \\
&= 1 \big/ f'(b)
\end{aligned}$$

and the proof is complete. □

Notice that the inverse function rule is another occasion when $\frac{dy}{dx}$ behaves like a fraction.

For an example of the inverse function rule, consider the function

$$f(x) = (x + 1)^7$$

This function is one-to-one, and hence has an inverse, $g(x)$. Let's use the inverse function rule to determine $g'(128)$.

By the chain rule, $f'(x) = 7(x + 1)^6$. Since $f(1) = 2^7 = 128$, we have $g(128) = 1$. Thus,

$$g'(128) = \frac{1}{f'(1)} = \frac{1}{7(2^6)} = \frac{1}{448}$$

Of course, for this example we can also solve the equation

$$y = (x + 1)^7$$

to give the equation for $x = g(y)$:

$$x = y^{1/7} - 1$$

Then,

$$\frac{dx}{dy} = \frac{1}{7} x^{-6/7} = \frac{1}{7x^{6/7}}$$

so, putting $y = 1$, we obtain the result $\frac{1}{448}$, as before.

3.2.3 The Product Rule

If $y = uv$, where u and v are both functions of x, then the *product rule* gives $\frac{dy}{dx}$ in terms of $\frac{du}{dx}$ and $\frac{dv}{dx}$.

Theorem (The Product Rule). If $y = uv$, where u and v are both differentiable functions of x, then y is a differentiable function of x and

$$\frac{dy}{dx} = u\frac{dv}{dx} + v\frac{du}{dx}$$

Note the symmetry in this rule. You take the first function times the derivative of the second plus the second function times the derivative of the first.

In functional notation, if $f = uv$, where u and v are both differentiable functions of x, then f is differentiable and

$$f' = uv' + vu' \qquad \blacksquare$$

For an example, let's use the product rule to find $\frac{dy}{dx}$ where

$$y = (x^3 - 4x + 1)(5x^2 - 3x - 2)$$

Taking $u = x^3 - 4x + 1$, $v = 5x^2 - 3x - 2$,

$$\frac{dy}{dx} = (x^3 - 4x + 1)(10x - 3) + (5x^2 - 3x - 2)(3x^2 - 4)$$

If necessary, we could simplify this result by collecting terms, but the differentiation is complete.

The proof of the product rule involves the use of an ingenious algebraic manipulation to express the definition of the derivative of f in terms of the definitions of the derivatives of u and v.

Proof of the Product Rule

By definition of the derivative,

$$f'(x) = \lim_{h \to 0} \frac{u(x+h)v(x+h) - u(x)v(x)}{h}$$

The idea is to try to transform this expression into a form where we have the definitions of the derivatives $u'(x)$ and $v'(x)$. If we subtract and then add back $u(x+h)v(x)$ in the numerator, we can rewrite the above limit as

$$\lim_{h \to 0} \frac{u(x{+}h)v(x{+}h) - u(x{+}h)v(x) + u(x{+}h)v(x) - u(x)v(x)}{h}$$

Rearranging the fraction, we can rewrite the limit as

$$\lim_{h \to 0} \left[u(x+h)\, \frac{v(x+h) - v(x)}{h} + v(x)\, \frac{u(x+h) - u(x)}{h} \right]$$

Evaluating this limit (treating each of the four constituents separately), we obtain

$$f'(x) = u(x)v'(x) + v(x)u'(x) \qquad \square$$

3.2.4 The Quotient Rule

If $y = u/v$, where u and v are both functions of x, then the *quotient rule* gives $\frac{dy}{dx}$ in terms of $\frac{du}{dx}$ and $\frac{dv}{dx}$.

Theorem (The Quotient Rule). If $y = \frac{u}{v}$, where u and v are both differentiable functions of x, then y is a differentiable function of x and

$$\frac{dy}{dx} = \left[v\frac{du}{dx} - u\frac{dv}{dx}\right] \Big/ v^2$$

In functional notation, if $f = \frac{u}{v}$, where u and v are differentiable functions of x, then f is differentiable, and

$$f' = \frac{vu' - uv'}{v^2} \qquad \blacksquare$$

For an illustration of the quotient rule, let's use it to find $\frac{dy}{dx}$ where

$$y = \frac{x^2 - 1}{x^2 + 1}$$

Taking $u = x^2 - 1$, $v = x^2 + 1$,

$$\frac{dy}{dx} = \frac{(x^2 + 1)(2x) - (x^2 - 1)(2x)}{(x^2 + 1)^2}$$

which simplifies to

$$\frac{dy}{dx} = \frac{4x}{(x^2 + 1)^2}$$

Proof of the Quotient Rule

As with the product rule, the proof of the quotient rule involves the use of an ingenious algebraic manipulation to express the definition of the derivative of f in terms of the definitions of the derivatives of u and v.

$$\begin{aligned} f'(x) &= \lim_{h \to 0} \left[\frac{u(x+h)}{v(x+h)} - \frac{u(x)}{v(x)}\right] \Big/ h \\ &= \lim_{h \to 0} \frac{v(x)u(x+h) - u(x)v(x+h)}{v(x+h)v(x)h} \end{aligned}$$

If we subtract and then add back $u(x)v(x)$ in the numerator, we can rewrite the limit as

$$\lim_{h\to 0}\frac{v(x)u(x+h)-u(x)v(x)+u(x)v(x)-u(x)v(x+h)}{v(x+h)v(x)h}$$

which rearranges as

$$\lim_{h\to 0}\left[v(x)\frac{u(x+h)-u(x)}{h}-u(x)\frac{v(x+h)-v(x)}{h}\right]\bigg/v(x+h)v(x)$$

Evaluating the limit (treating each of the five constituents separately), we obtain

$$f'(x)=[v(x)u'(x)-u(x)v'(x)]/v^2(x) \qquad \square$$

3.2.5 Examples

Example 1

Find $\dfrac{dy}{dx}$, where

$$y=(x^2+1)^5(x^3+1)^7$$

Solution

Let $u=(x^2+1)^5$, $v=(x^3+1)^7$. Then, using the chain rule,

$$\begin{aligned}\frac{du}{dx}&=5(x^2+1)^4(2x)=10x(x^2+1)^4\\ \frac{dv}{dx}&=7(x^3+1)^6(3x^2)=21x^2(x^3+1)^6\end{aligned}$$

Hence, using the product rule,

$$\begin{aligned}\frac{dy}{dx}&=u\frac{dv}{dx}+v\frac{du}{dx}\\ &=(x^2+1)^5\cdot 21x^2(x^3+1)^6+(x^3+1)^7\cdot 10x(x^2+1)^4\\ &=x(x^2+1)^4(x^3+1)^6[21x(x^2+1)+10(x^3+1)]\\ &=x(x^2+1)^4(x^3+1)^6(31x^3+21x+10)\end{aligned}$$

Example 2

Find $\dfrac{dy}{dx}$, where

$$y=\frac{x}{(x^2+1)^3}$$

Solution

Let $u = x$, $v = (x^2 + 1)^3$. Then, using the chain rule for the function v,

$$\frac{du}{dx} = 1$$

$$\frac{dv}{dx} = 3(x^2 + 1)^2 2x = 6x(x^2 + 1)^2$$

Hence, using the quotient rule,

$$\begin{aligned}
\frac{dy}{dx} &= \left[v\frac{du}{dx} - u\frac{dv}{dx}\right] \Big/ v^2 \\
&= \frac{(x^2 + 1)^3 \cdot 1 - x \cdot 6x(x^2 + 1)^2}{(x^2 + 1)^6} \\
&= \frac{(x^2 + 1)^2[(x^2 + 1) - 6x^2]}{(x^2 + 1)^6} \\
&= \frac{(x^2 + 1)^2(1 - 5x^2)}{(x^2 + 1)^6} \\
&= \frac{1 - 5x^2}{(x^2 + 1)^4}
\end{aligned}$$

3.2.6 Problems

1. Find $f'(x)$ for each of the following functions:

(a) $f(x) = (3x^3 + 2x^2 + x + 1)^9$

(b) $f(x) = \left(\dfrac{x - 1}{x + 1}\right)^2$

(c) $f(x) = \sqrt{\dfrac{x - 1}{x + 1}}$

(d) $f(x) = (x^3 + x^2 + x + 1)(x^5 + x^4 + 1)$

(e) $f(x) = \dfrac{x^6 + x^4 + x^2}{x^5 + x^3 + x + 1}$

(f) $f(x) = x\sqrt{1 - x^2}$

(g) $f(x) = \dfrac{x}{\sqrt{x^2 + 1}}$

2. Find $\dfrac{dy}{dx}$, when $x = \sqrt{y} + 3$.

3. Find $\dfrac{dy}{dx}$, when $x = y\sqrt{1 - y^2}$.

4. Calculate the slope of the curve

$$x = y^2 - 9y$$

where it crosses the x-axis.

3.3 TRIG FUNCTIONS

In this section we'll prove the following basic rules for differentiating trig functions:

$$\frac{d}{dx}(\sin x) = \cos x \qquad \frac{d}{dx}(\cos x) = -\sin x$$

$$\frac{d}{dx}(\tan x) = \sec^2 x \qquad \frac{d}{dx}(\cot x) = -\csc^2 x$$

$$\frac{d}{dx}(\sec x) = \sec x \tan x \qquad \frac{d}{dx}(\csc x) = -\csc x \cot x$$

By combining these results with the chain rule, we obtain more general rules, such as

$$\frac{d}{dx}\sin ax = a\cos ax$$

$$\frac{d}{dx}\cos ax = -a\sin ax$$

etc.

Theorem. If $f(x) = \sin x$, then $f'(x) = \cos x$.

Proof

$$f'(x) = \lim_{h\to 0}\frac{f(x+h) - f(x)}{h}$$

(By definition of the derivative.)

$$= \lim_{h\to 0}\frac{\sin(x+h) - \sin x}{h}$$

$$= \lim_{h\to 0}\frac{\sin x\cos h + \cos x\sin h - \sin x}{h}$$

(Using the trig expansion

$\sin(A + B) = \sin A\cos B + \cos A\sin B$.)

$$= \lim_{h\to 0}\frac{\sin x(\cos h - 1) + \cos x\sin h}{h}$$

(Rearranging the numerator.)

$$= \lim_{h\to 0} \frac{\sin x(\cos h - 1)}{h} + \lim_{h\to 0} \frac{\cos x \, \sin h}{h}$$

(The limit of a sum is the sum of the limits.)

$$= \sin x \lim_{h\to 0} \frac{\cos h - 1}{h} + \cos x \lim_{h\to 0} \frac{\sin h}{h}$$

(The limit of a product is the product of the limits.)

$$= (\sin x)(0) + (\cos x)(1)$$

(Using the theorems on trig limits in Section 2.2.6.)

$$= \cos x \qquad \square$$

Theorem. If $f(x) = \cos x$, then $f'(x) = -\sin x$.

Proof

We can write $f(x)$ as

$$f(x) = \sin\left(x + \frac{\pi}{2}\right)$$

Then, by the chain rule,

$$f'(x) = \cos\left(x + \frac{\pi}{2}\right) = -\sin x$$

and we are done. $\square$

Theorem. If $f(x) = \tan x$, then $f'(x) = \sec^2 x$.

Proof

Since $\tan x = \dfrac{\sin x}{\cos x}$, we can use the quotient rule.

$$\begin{aligned}
\frac{d}{dx}\tan x &= \frac{d}{dx}\left(\frac{\sin x}{\cos x}\right) \\
&= \frac{(\cos x)(\cos x) - (\sin x)(-\sin x)}{\cos^2 x} \\
&= \frac{\cos^2 x + \sin^2 x}{\cos^2 x} \\
&= \frac{1}{\cos^2 x} \quad (\text{since } \cos^2 x + \sin^2 x = 1) \\
&= \sec^2 x \qquad \square
\end{aligned}$$

Theorem. If $f(x) = \cot x$, then $f'(x) = -\csc^2 x$.

Proof

$$\frac{d}{dx}\cot x = \frac{d}{dx}(\tan x)^{-1}$$

$$
\begin{aligned}
&= -(\tan x)^{-2}\frac{d}{dx}(\tan x) \qquad \text{(By the chain rule.)}\\
&= -\frac{1}{\tan^2 x}\sec^2 x\\
&= -\frac{\cos^2 x}{\sin^2 x}\frac{1}{\cos^2 x}\\
&= -\frac{1}{\sin^2 x}\\
&= -\csc^2 x \qquad \square
\end{aligned}
$$

Theorem. If $f(x) = \sec x$, then $f'(x) = \sec x \tan x$.

Proof

$$
\begin{aligned}
\frac{d}{dx}\sec x &= \frac{d}{dx}(\cos x)^{-1}\\
&= -(\cos x)^{-2}\frac{d}{dx}(\cos x) \qquad \text{(By the chain rule.)}\\
&= \frac{1}{\cos^2 x}\sin x\\
&= \sec x \tan x \qquad \square
\end{aligned}
$$

Theorem. If $f(x) = \csc x$, then $f'(x) = -\csc x \cot x$.

Proof

$$
\begin{aligned}
\frac{d}{dx}\csc x &= \frac{d}{dx}(\sin x)^{-1}\\
&= -(\sin x)^{-2}\frac{d}{dx}(\sin x) \qquad \text{(By the chain rule.)}\\
&= -\frac{1}{\sin^2 x}\cos x\\
&= -\csc x \cot x \qquad \square
\end{aligned}
$$

3.3.1 Examples

Differentiate the following functions:

(a) $f(x) = 3\sin 5x + 2\cos 3x$

(b) $f(x) = \sin^3 x$

(c) $f(x) = \sqrt{\tan x}$

(d) $f(x) = \sin x \tan x$

(e) $f(x) = \sin(x^3)$

Solution

(a) $f'(x) = 15\cos 5x - 6\sin 3x$

(b) By the chain rule,

$$f'(x) = 3\sin^2 x \cos x$$

(c) Write the function as $f(x) = (\tan x)^{1/2}$. Then, applying the chain rule,

$$f'(x) = \tfrac{1}{2}(\tan x)^{-1/2}\sec^2 x = \frac{\sec^2 x}{2\sqrt{\tan x}}$$

(d) By the product rule,

$$f'(x) = \sin x \sec^2 x + \cos x \tan x$$

We can rewrite this as

$$f'(x) = \sin x \frac{1}{\cos^2 x} + \cos x \frac{\sin x}{\cos x} = \sec x \tan x + \sin x$$

(e) By the chain rule,

$$f'(x) = 3x^2 \cos(x^3)$$

3.3.2 Problems

1. Differentiate the following functions:

   (a) $f(x) = 5\csc 3x - 2\sec 7x$

   (b) $f(x) = \sec^5 x$

   (c) $f(x) = \sqrt{\cot x}$

   (d) $f(x) = \sec x \tan x$

   (e) $f(x) = \sin\sqrt{x}$

   (f) $f(x) = \tan(3x^4)$

   (g) $f(x) = \dfrac{x}{1 + \tan x}$

3.4 EXPONENTIALS AND LOGS

Let a be a positive constant. In Topic 1, we saw how the function a^r is defined for any rational exponent r. There is an alternative definition of the function a^x that works for any real number x. The two definitions give exactly the same values for rational exponenents. In fact, there are several ways to define a^x, though all require advanced concepts of calculus. We'll give one definition in Topic 6. For our present purposes, all we need to know is that the function a^x is continuous, a fact we assume throughout.

Theorem. The derivative of a^x is a constant multiple of a^x.

Proof

$$\begin{aligned}\frac{d}{dx}[a^x] &= \lim_{h\to 0}\frac{a^{x+h}-a^x}{h} = \lim_{h\to 0}\frac{a^x a^h - a^x}{h}\\ &= \lim_{h\to 0} a^x\,\frac{a^h-1}{h} = a^x \lim_{h\to 0}\frac{a^h-1}{h}\\ &= Z_a a^x\end{aligned}$$

where Z_a denotes the constant

$$Z_a = \lim_{h\to 0}\frac{a^h-1}{h}$$

(We assume that this limit exists, which it does.) The proof is complete. □

Let's look at the coefficient that arose in the preceding proof:

$$Z_a = \lim_{h\to 0}\frac{a^h-1}{h}$$

We shall presently obtain an alternative characterization of Z_a. In the meantime, we define the mathematical constant e to be the number such that the coefficient $Z_e = 1$.

[To convince yourself that it is reasonable to expect such a number e to exist, use a calculator to confirm that $Z_2 < 1$ and $Z_3 > 1$. Assuming that Z_x varies continuously with x (it does), there must then be some point e between 2 and 3 such that the curve $y = Z_x$ crosses the line $y = 1$.]

Since $Z_e = 1$, we have chosen e to satisfy the equation

$$\lim_{h\to 0}\frac{e^h-1}{h} = 1$$

As noted above, $Z_2 < 1$ and $Z_3 > 1$, so $2 < e < 3$. We show presently that $e = 2.71828\ldots$.

The function e^x (for the particular constant e) is generally referred to as "*the* exponential function." By the theorem we have just proved,

$$\frac{d}{dx}e^x = e^x$$

By the chain rule, we have

Theorem. For any constant k,

$$\frac{d}{dx}e^{kx} = ke^{kx} \qquad \blacksquare$$

An alternative way to define the real number e is

$$e = \lim_{h \to 0}(1 + h)^{1/h}$$

The number e is irrational, approximately 2.71828. Using a calculator, try working out a few values of the above expression for some small values of h, say $h = 0.1$, 0.01, 0.001, 0.0001. (The latter gives a value for e correct to three decimal places.)

Like π, the number e is a special constant in mathematics that arises in many different places.

The function $\log_e x$ is called the *natural logarithm*, often denoted by $\ln x$. The natural logarithm function is the inverse to the function e^x:

$$y = \ln x \iff x = e^y$$

Figure 3.5 indicates how the graphs of e^x and $\ln x$ are related. You can obtain values of $\ln x$ by reading the graph of e^x from the y-axis to the x-axis.

The derivative of the natural logarithm function is easy to remember:

Theorem.

$$\frac{d}{dx}\ln x = \frac{1}{x}$$

Proof

We start with the theorem

$$\frac{d}{dx}e^x = e^x$$

and use the inverse function rule. Let

$$y = \ln x$$

Then $x = e^y$, so

$$\frac{dx}{dy} = e^y$$

By the inverse function rule,

$$\frac{dy}{dx} = 1\bigg/\frac{dx}{dy} = 1/e^y = 1/x$$

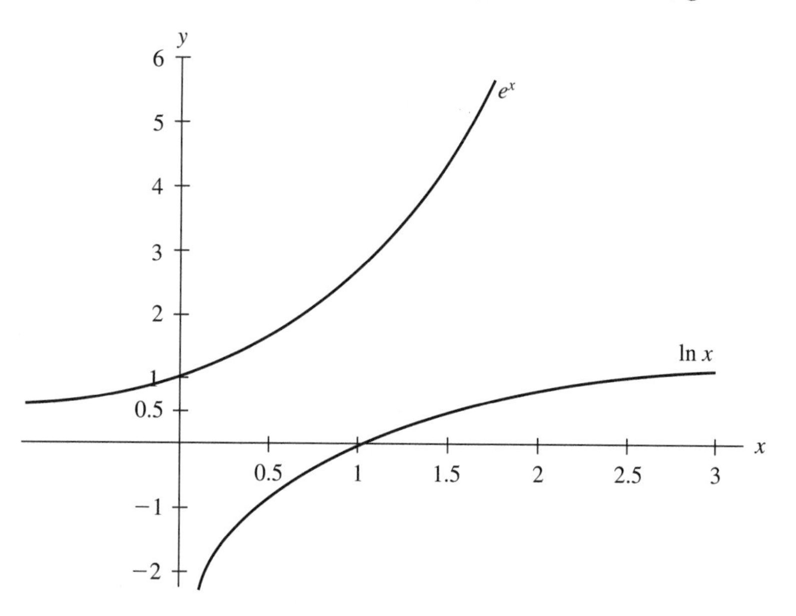

FIGURE 3.5 *The exponential and natural logarithm functions*

Thus,

$$\frac{d}{dx} \ln x = \frac{1}{x} \qquad \square$$

We have seen that for any positive constant a, $\dfrac{d}{dx}a^x = Z_a a^x$, where Z_a is the number

$$Z_a = \lim_{h \to 0} \frac{a^h - 1}{h}$$

We can now find another characterization of the coefficient Z_a: $Z_a = \ln a$.

Theorem. If $y = a^x$, then $\dfrac{dy}{dx} = (\ln a)a^x$.

Proof

Taking logs,

$$\ln y = x \ln a$$

Differentiating both sides of this equation with respect to x, using the chain rule,

$$\frac{d}{dy}(\ln y)\frac{dy}{dx} = \frac{d}{dx}(x \ln a)$$

That is,

$$\frac{1}{y}\frac{dy}{dx} = \ln a$$

Thus,

$$\frac{dy}{dx} = y \ln a = a^x \ln a$$

and we're done. □

Since we proved earlier that $\frac{d}{dx}(a^x) = Z_a a^x$, comparison with the above theorem shows that

$$Z_a = \ln a$$

3.4.1 Examples

✎ *Example 1*

Let $a > 0, a \neq 1$. Show that

$$\frac{d}{dx}(\log_a x) = \frac{1}{x}\log_a e = \frac{1}{(\ln a)x}$$

Solution

Let $y = \log_a x$. Thus, $x = a^y$.

Differentiating the last equation with respect to y,

$$\frac{dx}{dy} = (\ln a)a^y = (\ln a)x$$

By the inverse function rule,

$$\frac{dy}{dx} = \frac{1}{(\ln a)x}$$

This is part of the answer. Since $a = e^{\ln a}$, taking logarithms to base a, we have,

$$1 = \log_a a = \log_a(e^{\ln a}) = (\ln a)\log_a e$$

so

$$\log_a e = \frac{1}{\ln a}$$

and the other part of the result follows immediately.

Example 2

Use the chain rule and the properties of e^x and $\ln x$ to show that for any real number r,

$$\frac{d}{dx}[x^r] = rx^{r-1}$$

for $x > 0$.

(This is the fundamental result on differentiating a power of x that we stated as a theorem in Section 3.1.3, but did not prove at the time.)

Solution

Let $y = x^r$. Then, since $x = e^{\ln x}$, we have

$$y = \left(e^{\ln x}\right)^r = e^{r \ln x}$$

So, if we set $z = r \ln x$, we have $y = e^z$, and hence

$$\frac{dz}{dx} = \frac{r}{x},$$

$$\frac{dy}{dz} = e^z$$

Thus, using the chain rule:

$$\begin{aligned}\frac{dy}{dx} &= \frac{dy}{dz}\frac{dz}{dx} = \frac{r}{x}e^z = \frac{r}{x}y \\ &= \frac{r}{x}x^r = rx^{r-1}\end{aligned}$$

We're done.

Example 3

Find the derivative of

$$y = \sin 2\pi x + \ln \sqrt{1 - 4\pi x^2}$$

Solution

Using standard properties of logarithms, we can reformulate the function as:

$$y = \sin 2\pi x + \tfrac{1}{2}\ln(1 - 4\pi x^2)$$

Then, using the chain rule, together with the fact that the derivative of $\ln x$ is $1/x$, we have

$$\begin{aligned}\frac{dy}{dx} &= 2\pi \cos 2\pi x + \tfrac{1}{2}\left[\frac{1}{1 - 4\pi x^2}(-8\pi x)\right] \\ &= 2\pi \cos 2\pi x - \frac{4\pi x}{1 - 4\pi x^2}\end{aligned}$$

3.4.2 Problems

1. Differentiate the following functions:

(a) $f(x) = 5e^{7x}$

(b) $f(x) = e^{x^2}$

(c) $f(x) = \ln(5x^3)$

(d) $f(x) = x^2 e^{\pi x}$

(e) $f(x) = \dfrac{e^x + e^{-x}}{e^x - e^{-x}}$

(f) $f(x) = e^{\sqrt{x}} + \ln\sqrt{x}$

(g) $f(x) = e^x(\sin x + \cos x)$

(h) $f(x) = e^{\sin x}$

(i) $f(x) = x\ln(1 + x^2)$

(j) $f(x) = x(\ln x - 1)$

3.5 HIGHER-ORDER DERIVATIVES

3.5.1 Definitions

Differentiation takes one function $f(x)$ and produces a second function $f'(x)$. The first function has to be differentiable. The derivative $f'(x)$ may or may not be differentiable.

If $f'(x)$ is differentiable, we can differentiate it to produce a third function $f''(x)$, the derivative of $f'(x)$. The derivative of the derivative of $f(x)$ is generally referred to as the *second derivative* of $f(x)$.

If the second derivative $f''(x)$ of $f(x)$ is differentiable, differentiation of $f''(x)$ produces the *third derivative* of $f(x)$, denoted by $f'''(x)$. Similarly for the fourth derivative, etc.

To avoid a proliferation of superscripted primes, the *higher-order derivatives*

$$f''(x), \; f'''(x), \; f''''(x), \text{ etc.}$$

are usually written as

$$f^{(2)}(x), \; f^{(3)}(x), \; f^{(4)}(x), \text{ etc.}$$

Using dy/dx notation, the higher-order derivatives are written like this:

$$\frac{d^2y}{dx^2}, \frac{d^3y}{dx^3}, \frac{d^4y}{dx^4}, \ldots$$

3.5.2 Examples

Example 1

Find $f'(x),\ f''(x),\ f'''(x),\ f^{(4)}(x),\ f^{(5)}(x)$, where

$$f(x) = e^x + e^{-x}$$

Solution

We have:

$$\begin{aligned} f'(x) &= e^x - e^{-x} \\ f''(x) &= e^x + e^{-x} \\ f'''(x) &= e^x - e^{-x} \\ f^{(4)}(x) &= e^x + e^{-x} \\ f^{(5)}(x) &= e^x - e^{-x} \end{aligned}$$

Example 2

Find $f'(x),\ f''(x),\ f'''(x),\ f^{(4)}(x),\ f^{(5)}(x)$, where

$$f(x) = x^4$$

Solution

We have:

$$\begin{aligned} f'(x) &= 4x^3 \\ f''(x) &= 12x^2 \\ f'''(x) &= 24x \\ f^{(4)}(x) &= 24 \\ f^{(5)}(x) &= 0 \end{aligned}$$

Example 3

Find $\frac{dy}{dx},\ \frac{d^2y}{dx^2},\ \frac{d^3y}{dx^3},\ \frac{d^4y}{dx^4},\ \frac{d^5y}{dx^5}$, where

$$y = \sin 2x$$

Solution

We have:

$$\begin{aligned} \frac{dy}{dx} &= 2\cos 2x \\ \frac{d^2y}{dx^2} &= -4\sin 2x \\ \frac{d^3y}{dx^3} &= -8\cos 2x \end{aligned}$$

$$\frac{d^4y}{dx^4} = 16 \sin 2x$$
$$\frac{d^5y}{dx^5} = 32 \cos 2x$$

3.5.3 Problem

1. Write down a general formula for the nth derivative for each of the three functions in the above set of examples.

3.6 THE MEAN VALUE THEOREM

3.6.1 The Theorem

Suppose the function f is differentiable. Let a, b be points on the x-axis, $a < b$. The *mean slope* of the curve $y = f(x)$ between $x = a$ and $x = b$ is the net rise divided by the net run, namely

$$\frac{f(b) - f(a)}{b - a}$$

The *Mean Value Theorem* says that for some point x_0 between a and b, the slope of the curve at $x = x_0$ is exactly equal to the mean slope between a and b. The theorem is illustrated in Figure 3.6. The number x_0 is chosen so that the tangent to the curve at the point $(x_0, f(x_0))$ is parallel to the chord from $(a, f(a))$ to $(b, f(b))$.

Here is the formal statement of the Mean Value Theorem.

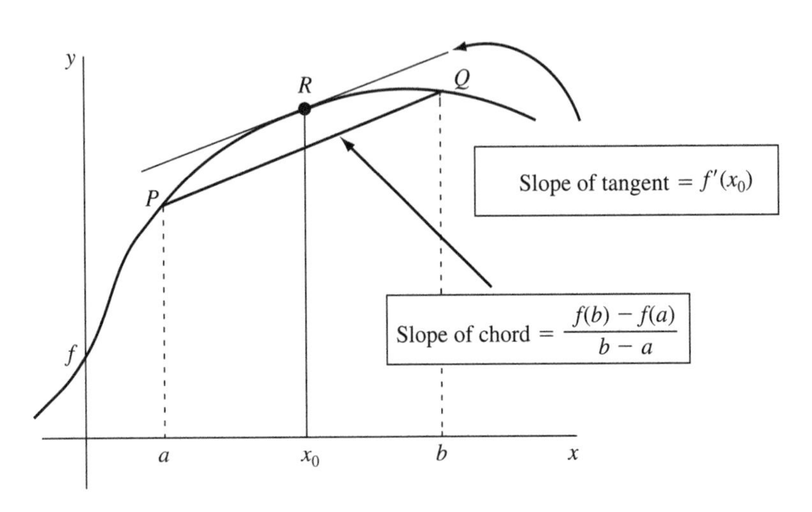

FIGURE 3.6 *The Mean Value Theorem*

Theorem (Mean Value Theorem). If $f(x)$ is continuous on the interval $[a, b]$ and differentiable on (a, b), then there is a point x_0 in (a, b) such that

$$f'(x_0) = \frac{f(b) - f(a)}{b - a} \qquad \blacksquare$$

See the problem set below for a proof of the mean value theorem, starting from a more basic result known as Rolle's theorem. Here is an intuitive explanation (not a formal proof) of the theorem. Imagine taking the straight line PQ in Figure 3.6 and translating it upward (keeping the slope fixed). The endpoints P and Q get closer together. In the final position before the line no longer intersects the curve, P and Q coincide and the line is tangent to the curve at that common point. That common point is the point R; at R, the slope of the curve is equal to the slope of the tangent line, which is the same as the slope of the original chord PQ, namely

$$\frac{f(b) - f(a)}{b - a}$$

The Mean Value Theorem provides a simple method for approximating particular values of certain functions. Here is the key idea.

MVT Approximation. If $f(x)$ is continuous on the interval $[a, b]$ and differentiable on (a, b), then for any x in (a, b) there is a point x_0 in (a, x) such that

$$f(x) = f(a) + (x - a)f'(x_0)$$

To prove this result, you just apply the Mean Value Theorem to f for the interval $[a, x]$.

3.6.2 Examples

Find a point x_0 as in the Mean Value Theorem for the function

$$f(x) = 2x^2 - 3x - 1$$

with $a = 1,\ b = 4$.

Solution

We have

$$f'(x) = 4x - 3,\ \ f(a) = f(1) = -2,\ \ f(b) = f(4) = 19$$

$$\frac{f(b) - f(a)}{b - a} = \frac{21}{3} = 7$$

We need to find x_0 to satisfy

$$4x_0 - 3 = 7$$

This can be solved to give $x_0 = 5/2$. Note that $x_0 \in (1, 4)$.

Example 2

Use the MVT approximation formula to show that

$$\sin x < x, \text{ for all } x > 0$$

Solution

Since $\sin x \leq 1$ for all x, the inequality is trivially valid if $x > 1$, so the only case we have to concern ourselves with is $0 < x \leq 1$.

Apply the MVT approximation formula to the function $f(x) = \sin x$ on the interval $[0, 1]$. For any x in the interval $(0, 1]$, there is an x_0 in $(0, x)$ such that

$$f(x) = f(0) + (x - 0)f'(x_0)$$

That is,

$$\sin x = \sin 0 + x \cos x_0 = x \cos x_0$$

Since $0 \leq x_0 < 1 < \pi/2$, we have $\cos x_0 < 1$, and hence $x \cos x_0 < x$. Thus $\sin x < x$, as required.

3.6.3 Problems

1. Find a point x_0 as in the Mean Value Theorem for the function

$$f(x) = x^3$$

with $a = 1, \ b = 3$.

2. Rolle's theorem says that if $f(x)$ is continuous on an interval $[a, b]$ and differentiable on (a, b), and if $f(a) = f(b)$, then $f'(x_0) = 0$ for some point x_0 in (a, b).

Here is an intuitive explanation (not a formal proof) of Rolle's theorem. If $f(x)$ is a constant function, the result is trivially true, since then $f'(x) = 0$ for all x in $[a, b]$. If $f(x)$ is not constant, then it cannot be increasing everywhere on (a, b), nor can it be decreasing everywhere on (a, b); otherwise we would not have $f(a) = f(b)$. So, for some point x_0 in (a, b), $f(x)$ must change from being increasing to decreasing, or vice versa. At that point, $f'(x_0) = 0$.

Use Rolle's theorem to prove the mean value theorem.

[*Hint.* Referring to Figure 3.6, the equation of the straight line PQ is

$$y = f(b) + K(x - b)$$

where

$$K = \frac{f(b) - f(a)}{b - a}$$

At any point x in the interval (a, b), the vertical distance from the line PQ to the curve is

$$F(x) = f(x) - f(b) - K(x - b)$$

Show that the function $F(x)$ satisfies the requirements of Rolle's theorem, and then apply the theorem to $F(x)$.]

dx

topic 4
Differentiation II

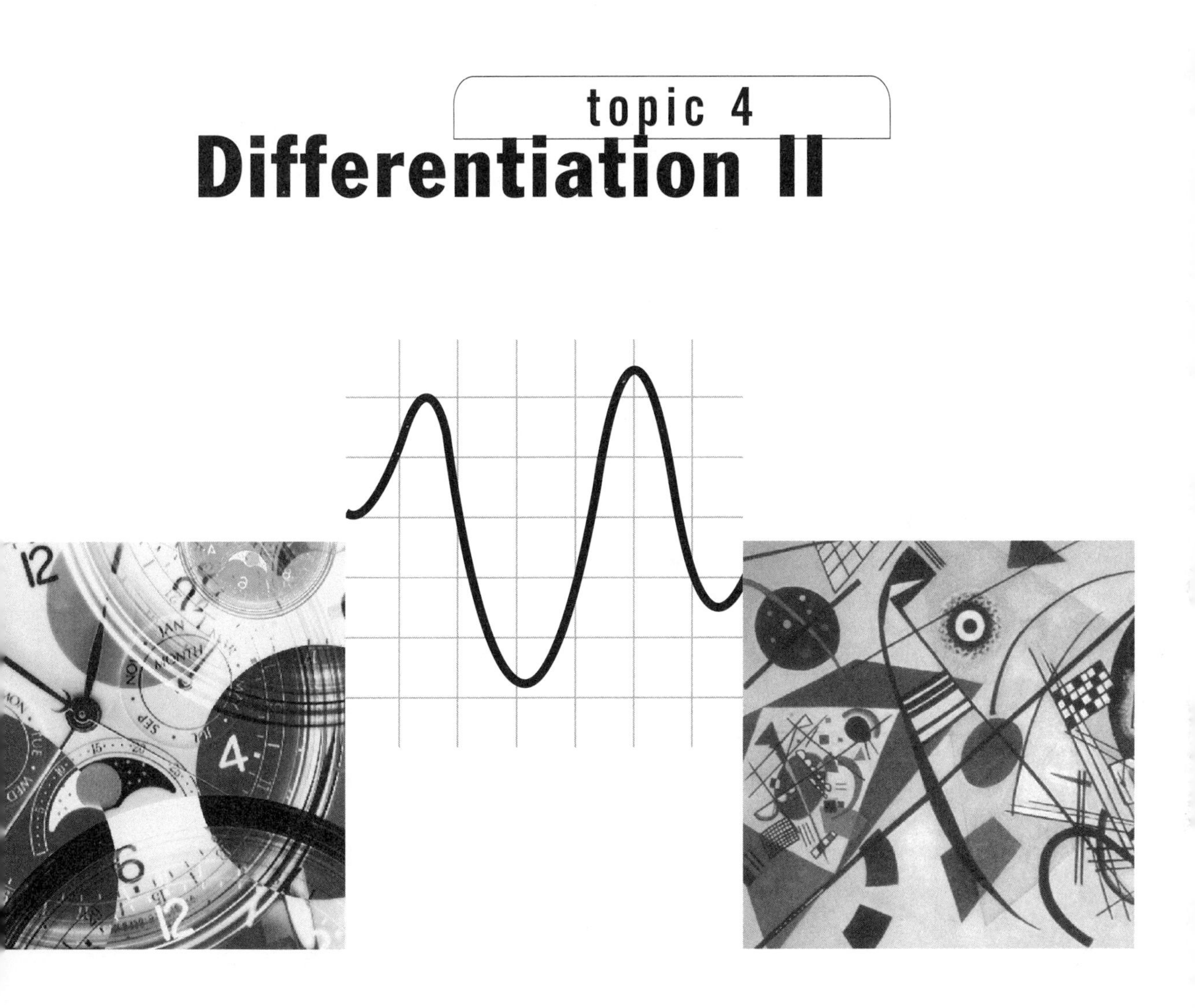

4.1 MAXIMA AND MINIMA

One of the most important applications of the differential calculus is the solution to problems of maximization and minimization. For example, such problems arise often in business, where the goal is to maximize profits and minimize costs.

Real-life maximization and minimization problems often involve two or more (independent) variables, and their solution is beyond the scope of this *Companion*. However, some maximization and minimization problems are single variable. For such problems, the calculus techniques we describe here may be applicable.

4.1.1 Definitions

A function $f(x)$ from the real line to the real line is said to have a *local* (or *relative*) *maximum* at a point c if $f(c) \geq f(x)$ for all x in some open interval containing c.

Similarly, $f(x)$ is said to have a *local* (or *relative*) *minimum* at a point c if $f(c) \leq f(x)$ for all x in some open interval containing c.

The following theorem indicates how you can use the differential calculus to determine maximum and minimum points.

Theorem. If $f(x)$ has a local maximum or a local minimum at c, and if $f'(c)$ exists, then $f'(c) = 0$.

Before we give the proof, notice that, geometrically, the result is obvious. If f has, say, a local maximum at c, then $f(c)$ must be greater than $f(x)$ for all x in some open interval $(c - \epsilon, c + \epsilon)$, and hence the slope of f must be zero at c, i.e., $f'(c) = 0$. (See Figure 4.1).

Proof

We consider first the case where f has a local maximum at c.

Since f has a local maximum at c, for some positive real number ϵ,

$$f(c) \geq f(c+h), \text{ whenever } |h| < \epsilon.$$

Hence,

$$f(c+h) - f(c) \leq 0, \text{ whenever } |h| < \epsilon. \tag{1}$$

Consider the case where $h > 0$ and $|h| < \epsilon$. Then dividing inequality (1) by the positive number h, we get

$$\frac{f(c+h) - f(c)}{h} \leq 0$$

This is true for any positive h less than ϵ. Hence

$$\lim_{h \to 0+} \frac{f(c+h) - f(c)}{h} \leq 0$$

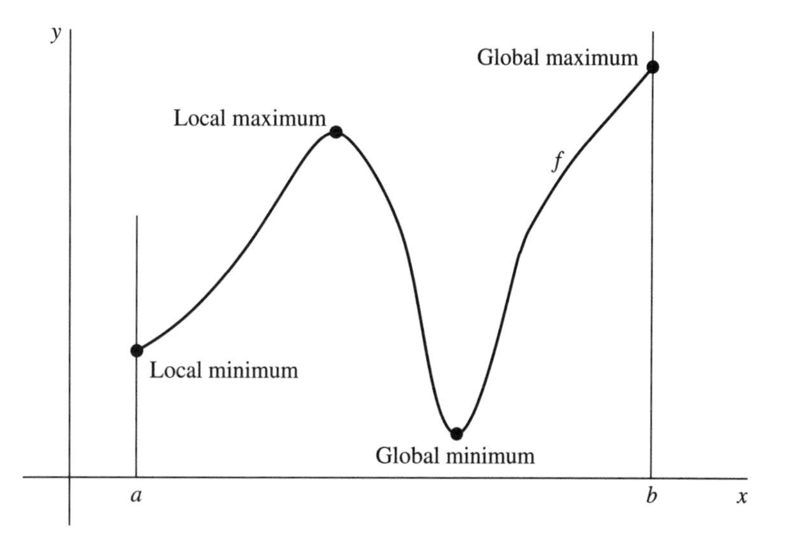

FIGURE 4.1 *Maxima and minima*

But since $f'(c)$ exists, we have

$$f'(c) = \lim_{h \to 0} \frac{f(c+h) - f(c)}{h} = \lim_{h \to 0+} \frac{f(c+h) - f(c)}{h}$$

Thus

$$f'(c) \leq 0$$

Now consider the case where $h < 0$ and $|h| < \epsilon$. Then, dividing inequality (1) by the negative number h, the direction of the inequality changes, to give

$$\frac{f(c+h) - f(c)}{h} \geq 0$$

Thus, in this case

$$f'(c) = \lim_{h \to 0} \frac{f(c+h) - f(c)}{h} = \lim_{h \to 0-} \frac{f(c+h) - f(c)}{h} \geq 0$$

We have now shown that $f'(c) \leq 0$ and $f'(c) \geq 0$. Thus, we must have $f'(c) = 0$, and the theorem is proved in this case.

The case for a local minimum is entirely analogous, so we won't give the details. □

If $f(c) \geq f(x)$ for all x in the domain of $f(x)$, we say that $f(x)$ has a *global* or *absolute maximum* at c.

If $f(c) \leq f(x)$ for all x in the domain of $f(x)$, we say that $f(x)$ has a *global* or *absolute minimum* at c.

Notice that for a function $f(x)$ whose domain is a closed interval $[a, b]$, a maximum or a minimum might occur at one of the endpoints, as in Figure 4.1. For a local or global maximum/minimum at an endpoint, the derivative $f'(x)$ might not be zero.

Local or global max/min points are sometimes referred to as *extreme points*.

There are several tests you can use to find local maxima and minima.

4.1.2 First Derivative Test (Version 1)

To determine the local maxima and minima of the differentiable function $f(x)$:

Step 1. Compute the derivative $f'(x)$.

Step 2. Solve the equation $f'(x) = 0$. The solutions are called *critical numbers* (or *critical points*).

Step 3. For each critical number c, compute $f(x)$ for points x close to c on the left (i.e., $x < c$) and on the right (i.e., $x > c$), making sure that there are no other critical numbers between c and the points x you choose.

Conclusions

- If $f(x) \geq f(c)$ for all values of x sufficiently close to c on the left and on the right, the function has a local minimum at c.
- If $f(x) \leq f(c)$ for all values of x sufficiently close to c on the left and on the right, the function has a local maximum at c.
- If $f(x) \geq f(c)$ for all x sufficiently close to c on the left and $f(x) \leq f(c)$ for all x sufficiently close to c on the right, or the other way round, then f has neither a local maximum nor a local minimum at c. It has what is called a *point of inflection* at c. A point of inflection is where the shape of the curve changes from concave up to concave down (see Section 4.1.5 for details), or vice versa. See Figure 4.2.

For example, suppose we want to find the maximum and minimum values of the function

$$f(x) = 2x^3 - 9x^2 + 12x + 1$$

Differentiating,

$$f'(x) = 6x^2 - 18x + 12$$

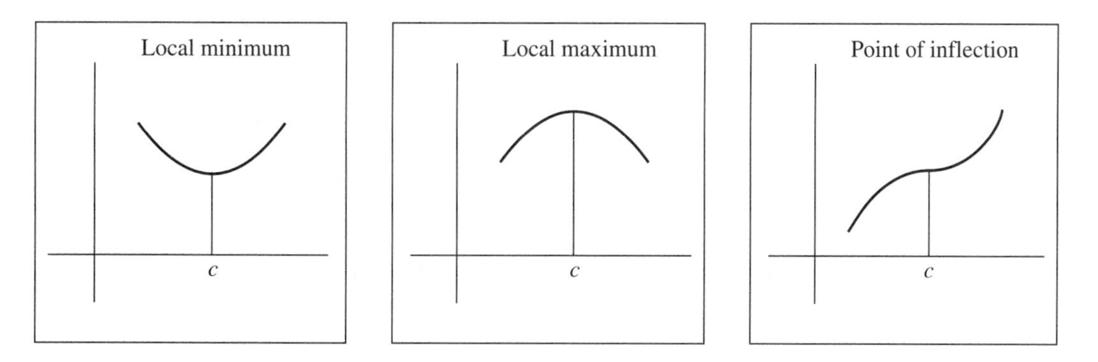

FIGURE 4.2 *Critical points*

This factors as

$$f'(x) = 6(x - 1)(x - 2)$$

The critical numbers are the points where $f'(x) = 0$, namely $x = 1, 2$. At the critical numbers, we have

$$f(1) = 6, \quad f(2) = 5$$

To see what kinds of points these are, we compute values of $f(x)$ close to the critical numbers. First, near to $x = 1$, we have

$$f(0.9) = 5.968, \quad f(1.1) = 5.972$$

These two values are both less than $f(1) = 6$, and it is clear that 0.9 and 1.1 are typical of points close to 1 in this regard, so we conclude that the function has a local maximum at $x = 1$.

Near to $x = 2$, we have

$$f(1.9) = 5.028, \quad f(2.1) = 5.032$$

These two values are both greater than $f(2) = 5$, and again 1.9 and 2.1 are typical of points close to 2 in this regard, so we conclude that the function has a local minimum at $x = 1$.

For a second example, suppose we want to determine the maximum and minimum values of the function

$$f(x) = x^3$$

Differentiating,

$$f'(x) = 3x^2$$

The critical numbers are the points where $f'(x) = 0$, namely $x = 0$. At the critical number, we have

$$f(0) = 0$$

To see what kind of point this is, we compute values of $f(x)$ close to the critical number.

$$f(-0.1) = -0.001, \quad f(0.1) = 0.001$$

The first value is less than $f(0) = 0$ and the second is greater, so we conclude that the function has a point of inflection at $x = 0$.

4.1.3 First Derivative Test (Version 2)

To determine the local maxima and minima of the differentiable function $f(x)$:

Step 1. Compute the derivative $f'(x)$.

Step 2. Solve the equation $f'(x) = 0$ to obtain the critical numbers.

Step 3. For each critical number c, determine the sign of $f'(x)$ for points x close to c on the left (i.e., $x < c$) and then for points x close to c on the right (i.e., $x > c$).

Conclusions

- If $f'(x)$ changes sign from negative to positive at c, the function has a local minimum at c.
- If $f'(x)$ changes sign from positive to negative at c, the function has a local maximum at c.
- If $f'(x)$ does not change sign at c, then f has neither a local maximum nor a local minimum at c; it has a point of inflection at c.

We'll use the previous two examples to illustrate this method.

First, we use the test to find the maximum and minimum values of the function

$$f(x) = 2x^3 - 9x^2 + 12x + 1$$

Differentiating,

$$f'(x) = 6x^2 - 18x + 12$$

This factors as

$$f'(x) = 6(x - 1)(x - 2)$$

The critical numbers are the points where $f'(x) = 0$, namely $x = 1, 2$. At the critical numbers, we have

$$f(1) = 6, \quad f(2) = 5$$

To see what kinds of points these are, we compute the sign of $f'(x)$ close to the critical numbers. First, near to $x = 1$, we have

$$f'(x) > 0, \text{ for } x < 1$$
$$f'(x) < 0, \text{ for } 1 < x < 2$$

so the function has a local maximum at $x = 1$.

Near to $x = 2$, we have

$$f'(x) < 0, \text{ for } 1 < x < 2$$
$$f'(x) > 0, \text{ for } x > 2$$

so the function has a local minimum at $x = 2$.

For the second example, we use the test to determine the maximum and minimum values of the function

$$f(x) = x^3$$

Differentiating,

$$f'(x) = 3x^2$$

The critical numbers are the points where $f'(x) = 0$, namely $x = 0$. At the critical number, we have

$$f(0) = 0$$

To see what kind of point this is, we compute the sign of $f'(x)$ close to the critical number:

$$f'(x) > 0, \text{ for all nonzero } x$$

Since $f'(x)$ does not change sign at $x = 0$, the function has a point of inflection at $x = 0$.

4.1.4 Second Derivative Test

Our third test for extreme points works for twice-differentiable functions. A function $f(x)$ is *twice differentiable* at a point $x = a$ if both $f'(a)$ and $f''(a)$ are defined. A function that is twice differentiable at all points in its domain is said to be a *twice-differentiable function.*

Geometrically, the idea behind the second derivative test is as follows. Suppose we have solved the equation $f'(x) = 0$ to find the critical numbers. For a critical number c, we want to know if f has a local maximum at c, a local minimum at c, or otherwise. If $f''(c) > 0$, that means that the slope $f'(x)$ of f is increasing at c. But we know that the slope is zero at c. Hence, the slope must be negative just to the left of c and positive just to the right of c. (Draw a picture!) Thus, the function has a local maximum at c.

On the other hand, if $f''(c) < 0$, that means that the slope $f'(x)$ of f is decreasing at c. Since the slope is zero at c, it follows that the slope must be positive just to the left of c and negative just to the right of c. Thus, the function has a local maximum at c. (Draw a picture!)

With this intuitive picture in our minds, let's take a look at the second derivative test.

To determine the local maxima and minima of the twice-differentiable function $f(x)$:

Step 1. Compute the derivatives $f'(x)$ and $f''(x)$.

Step 2. Solve the equation $f'(x) = 0$ to obtain the critical numbers.

Step 3. For each critical number c, compute $f''(c)$.

Conclusions

- If $f''(c) > 0$, the function has a local minimum at c.
- If $f''(c) < 0$, the function has a local maximum at c.
- If $f''(c) = 0$ and $f''(x)$ changes sign at c, then f has neither a local maximum nor a local minimum at c; it has a point of inflection at c.
- If $f''(c) = 0$ and $f''(x)$ does not change sign at c, then the test is inconclusive in this case, and further analysis is required.

We'll again use the same two examples to illustrate the second derivative test.

First, we use the test to find the maximum and minimum values of the function

$$f(x) = 2x^3 - 9x^2 + 12x + 1$$

Differentiating,

$$f'(x) = 6x^2 - 18x + 12$$

Differentiating again,

$$f''(x) = 12x - 18$$

The expression for $f'(x)$ factors as

$$f'(x) = 6(x-1)(x-2)$$

The critical numbers are the points where $f'(x) = 0$, namely $x = 1,\ 2$. At the critical numbers, we have

$$f(1) = 6,\ \ f(2) = 5$$

To see what kinds of points these are, we compute $f''(x)$ at the critical numbers. First, at $x = 1$, we have

$$f''(1) = -6$$

which is negative, so the function has a local maximum at $x = 1$. At $x = 2$, we have

$$f''(2) = 6$$

which is positive, so the function has a local minimum at $x = 2$.

For the second example, we use the second derivative test to determine the maximum and minimum values of the function

$$f(x) = x^3$$

Differentiating,

$$f'(x) = 3x^2$$

Differentiating again,

$$f''(x) = 6x$$

The critical numbers are the points where $f'(x) = 0$, namely $x = 0$. At the critical number, we have

$$f(0) = 0$$

To see what kind of point this is, we first compute $f''(x)$ at the critical number:

$$f''(0) = 0$$

Then we examine the sign of $f''(x)$ near the critical number:

$$f''(x) < 0, \quad \text{for } x < 0$$
$$f''(x) > 0, \quad \text{for } x > 0$$

Since $f''(x)$ changes sign at $x = 0$, the function has a point of inflection at $x = 0$.

The two functions used to illustrate the various tests were carefully chosen so that each test worked easily for each example. This enabled you to compare the different tests. In practice, things rarely work out so well, and you may well find yourself faced with a function for which not all the tests give an answer. So you do need to be familiar with all three tests.

4.1.5 Concavity

An arc of a curve $y = f(x)$ is said to be *concave upward* if, at each point on the arc, the arc lies above the tangent at that point.

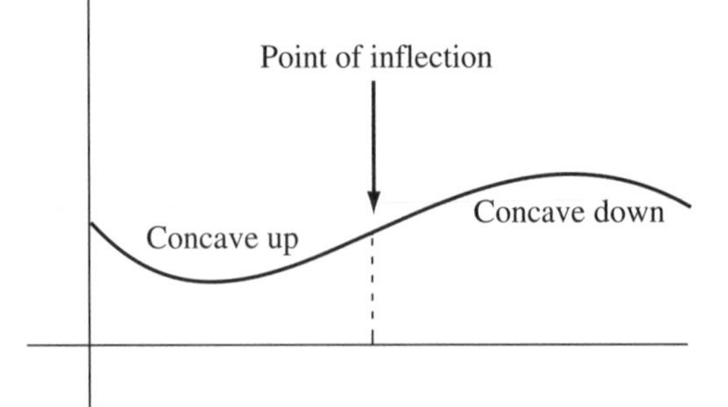

FIGURE 4.3 *Concavity*

Equivalently, the slope of the curve must be increasing as you go along the arc. See Figure 4.3.

Thus, if $f(x)$ is differentiable on the arc, the arc will be concave upward if and only if $f'(x)$ is increasing for all points on the arc.

Since an increasing differentiable function is one having a positive derivative, it follows that the arc of a twice-differentiable function will be concave upward if and only if

$$f''(x) > 0$$

for all points x on the arc.

An arc of a curve $y = f(x)$ is said to be *concave downward* if, at each point on the arc, the arc lies below the tangent at that point. Equivalently, the slope of the curve must be decreasing as you go along the arc. See Figure 4.3.

Thus, if $f(x)$ is differentiable on the arc, the arc will be concave downward if and only if $f'(x)$ is decreasing at all points on the arc.

If $f(x)$ is twice-differentiable on the arc, the curve will be concave downward if and only if

$$f''(x) < 0$$

for all points x on the arc.

A point on a curve where the curve changes from being concave upward to concave downward or vice versa is called a *point of inflection.* See Figure 4.3.

If $f(x)$ is twice-differentiable, then $f(x)$ has a point of inflection at $x = a$ if and only if

1. $f''(a) = 0$
2. $f''(x)$ changes sign at a.

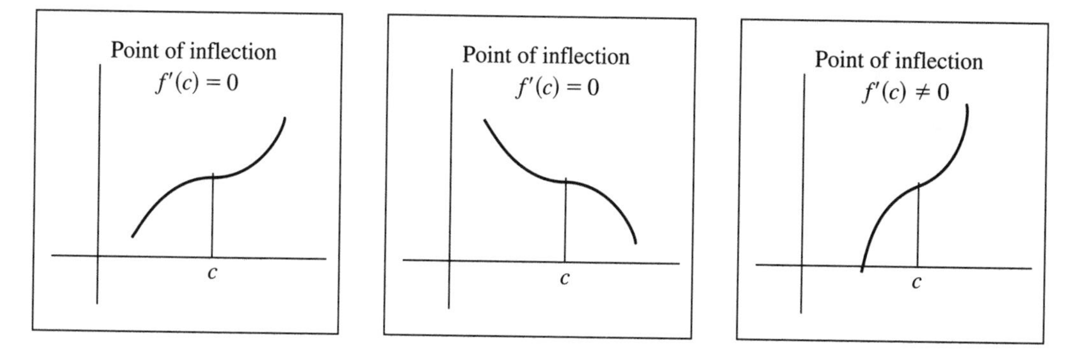

FIGURE 4.4 *Points of inflection*

Condition 2 means that there is an interval (l, a) such that $f''(x)$ is negative (respectively positive) on (l, a) and an interval (a, r) such that $f''(x)$ is positive (respectively negative) on (a, r).

Notice that it is possible for $f'(c) \neq 0$ at a point of inflection. In the three illustrations of points of inflection in Figure 4.4, the first two have $f'(c) = 0$ but the third has $f'(c) > 0$.

4.1.6 Examples

Example 1

Investigate the curve

$$y = f(x) = 3x^4 - 8x^3 - 6x^2 + 24x + 3$$

finding and classifying all critical points and determining the regions where the curve is concave up and concave down. The curve is shown in Figure 4.5.

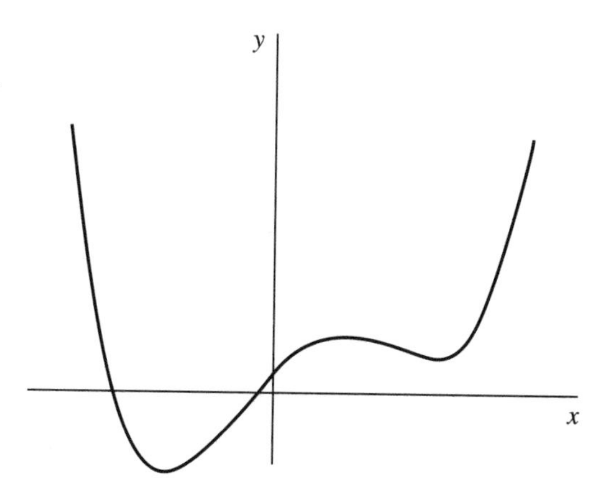

FIGURE 4.5 *Graph of* $y = 3x^4 - 8x^3 - 6x^2 + 24x + 3$

Solution

Differentiating,

$$f'(x) = 12x^3 - 24x^2 - 12x + 24$$
$$f''(x) = 36x^2 - 48x - 12$$

For a critical point, we have $f'(x) = 0$, i.e.,

$$12x^3 - 24x^2 - 12x + 24 = 0$$

This simplifies to

$$x^3 - 2x^2 - x + 2 = 0$$

which factors as

$$(x + 1)(x - 1)(x - 2) = 0$$

Thus the critical numbers are $x = -1,\ 1,\ 2.$

To classify the critical points, we'll use the second derivative test. For $x = -1$,

$$f''(-1) = 36 + 48 - 12 = 72 > 0$$

so there is a local minimum at $x = -1$.
For $x = 1$,

$$f''(1) = 36 - 48 - 12 = -24 < 0$$

so there is a local maximum at $x = 1$.
For $x = 2$,

$$f''(2) = 144 - 96 - 12 = 36 > 0$$

so there is a local minimum at $x = 2$.

To find the points of inflection, where the concavity of the curve changes, we solve $f''(x) = 0$, i.e.,

$$36x^2 - 48x - 12 = 0$$

This simplifies to

$$3x^2 - 4x - 1 = 0$$

Using the quadratic formula, this has the solutions

$$x = \frac{4 \pm \sqrt{16 + 12}}{6} = \frac{4 \pm \sqrt{28}}{6} = \frac{2 \pm \sqrt{7}}{3}$$

We examine the sign of $f''(x)$. To do this, we use the fact that the sign of a continuous function is constant between two consecutive zeros. Hence, to determine the sign of $f''(x)$ between the zeros of $f''(x)$, we simply need to evaluate the function at one point in each interval.

For $x < \dfrac{2-\sqrt{7}}{3}$, $f''(x) > 0$, so the curve is concave up in this region.

For $\dfrac{2-\sqrt{7}}{3} < x < \dfrac{2+\sqrt{7}}{3}$, $f''(x) < 0$, so the curve is concave down in this region.

For $x > \dfrac{2+\sqrt{7}}{3}$, $f''(x) > 0$, so the curve is concave up in this region.

Since the concavity changes at $x = \dfrac{2-\sqrt{7}}{3}$ and $x = \dfrac{2+\sqrt{7}}{3}$, the curve has a point of inflection at each of these two values of x.

✎ *Example 2*

Investigate the curve

$$y = f(x) = 1 + x^{2/3}$$

finding and classifying all critical points and determining the regions where the curve is concave up and concave down. The curve is shown in Figure 4.6.

Solution

Differentiating,

$$f'(x) = \frac{2}{3}x^{-1/3} = \frac{2}{3x^{1/3}}$$

$$f''(x) = -\frac{2}{3}\frac{1}{3}x^{-4/3} = -\frac{2}{9x^{4/3}}$$

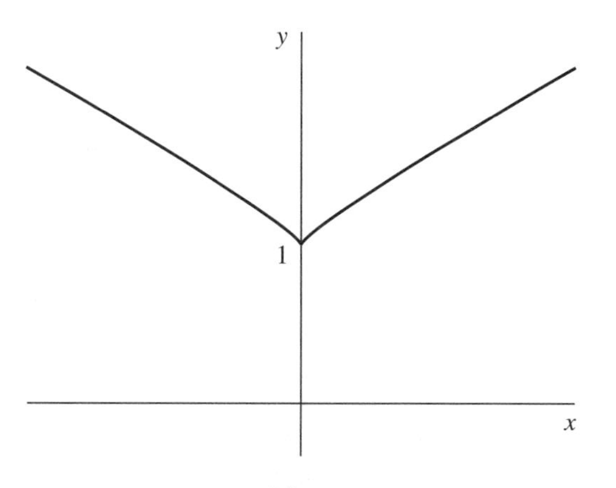

FIGURE 4.6 $y = 1 + x^{2/3}$

There are no numbers x for which $f'(x) = 0$. However, for $x = 0$, $f'(x)$ is undefined (i.e., it is infinite). We say that $x = 0$ is a critical number of a kind different from the ones we discussed earlier. Let's examine the behavior of the curve to the left and to the right of this critical point.

For $x < 0$, $f'(x) < 0$; for $x > 0$, $f'(x) > 0$. Thus the curve has a local minimum at $x = 0$.

To examine the concavity, we look at $f''(x)$.

For $x < 0$, $f''(x) < 0$, so the curve is concave downward for $x < 0$; for $x > 0$, $f''(x) < 0$, so the curve is also concave downward for $x > 0$.

4.1.7 Problems

1. Investigate the curve

$$y = x^4 + 2x^3 - 3x^2 - 4x + 4$$

finding and classifying all critical points and determining the regions where the curve is concave up and concave down. Graph the curve and compare your findings with the graph.

2. Investigate the curve

$$y = (x - 2)^{2/3}$$

finding and classifying all critical points and determining the regions where the curve is concave up and concave down. Graph the curve and compare your findings with the graph.

3. Investigate the curve

$$y = (x - 1)^{1/3}(x + 2)^{2/3}$$

finding and classifying all critical points and determining the regions where the curve is concave up and concave down. Graph the curve and compare your findings with the graph.

4. Use calculus to find the critical points of the curve

$$y = \frac{1}{x(x - 1)}$$

Then use a grapher to classify those critical points and to determine approximately the regions where the curve is concave up and concave down.

4.2 RATES OF CHANGE

Many applications of calculus involve functions of time, where some quantity u varies with time t:

$$u = f(t)$$

Such applications do not involve any different techniques of calculus. The distinction is in how we interpret the results. For example, the derivative $f'(t)$ represents the rate of change of u with respect to t. If u is the distance traveled by an object in time t, then $f'(t)$ is the speed of the object at time t and $f''(t)$ is the acceleration at time t.

Though detailed applications of calculus are outside the scope of the *Electronic Companion*, we give a brief overview of some common applications to rates of change.

4.2.1 Motion

You can apply the methods of calculus to situations involving motion whenever you can write down an equation that connects distance with time. There are mathematical techniques that you can use to produce a polynomial equation to fit any realistic set of data for some motion. Thus, one way to study the motion of some object is to tabulate distance against time, find a polynomial that fits the data, and then use calculus to examine the polynomial. Let's look at a fictional example.

A tourist with a video camera captures on tape a flying object (a UFO) in the sky over the New Mexico desert. An analysis of the tape by NASA scientists shows that the distance s in feet traveled by the object over a ten-second period, measured in seconds, is as follows:

t	1	2	3	4	5	6	7	8	9	10
s	29	50	82	145	108	408	640	956	1380	1915

A NASA mathematician examines the figures and produces a cubic equation that fits the data remarkably well:

$$s = 2.5t^3 - 9t^2 + 31t + 5$$

Differentiating this equation gives formulas for the speed and acceleration at each time of the flight:

$$\text{Speed}: \quad \frac{ds}{dt} = 7.5t^2 - 18t + 31$$

$$\text{Acceleration}: \quad \frac{d^2s}{dt^2} = 15t - 18$$

Thus, at $t = 1$, the UFO was traveling at around 20.5 feet per second, with little acceleration (the formula gives -3 ft/sec^2). The acceleration was increasing linearly throughout the period observed. At time $t = 10$, the speed of the UFO was around 601 ft/sec and its acceleration was around 132 ft/sec^2.

Graph the three functions of t: s, $\frac{ds}{dt}$, and $\frac{d^2s}{dt^2}$.

4.2.2 Calculus in Economics

Methods of calculus can often be used in the business world to maximize profits or to plan expansion. Here is the general idea.

Economists define the *cost function* $C(q)$ to be the cost of producing a quantity q of some product. The *revenue function* $R(q)$ is the total revenue received by selling quantity q.
If $p(q)$ is the unit price when quantity q is sold, then

$$R(q) = p(q) \times q$$

If $p(q)$ is constant, then $R(q)$ is a straight line. In many cases, however, $p(q)$ decreases as q increases. (The more you buy, the less the unit price.) The figure of most interest to the manufacturer is given by the profit function

$$\pi(q) = R(q) - C(q)$$

In the case where the product can be measured by a continuous variable, say the manufacture of a drug, a household or industrial chemical, oil or gas, soft drinks or liquor, then it may be possible to use the methods of calculus. For instance, if $C(q)$ and $R(q)$ can be expressed as differentiable functions, then it is easy to compute the quantity of product that will maximize profit. For a maximum profit,

$$\pi'(q) = 0$$

This will occur when

$$R'(q) = C'(q)$$

Even in cases where q is not a continuous variable, if the number of items manufactured and sold is sufficiently large that a single unit is 'negligibly small', then the variable q may be treated as if it were continuous, again enabling you to use methods of calculus.

For example, suppose you are the manager of an airline and you want to decide whether to introduce an additional flight. If your aim is to maximize profits, you want to know how your profit will change going from 100 flights to 101. This means you need to know how $C(q)$ and $R(q)$ will change if q increases from 100 to 101.

If you currently operate 100 flights, the increased cost of putting on one more flight is

$$C(101) - C(100)$$

The rate of change of your cost is

$$\frac{C(101) - C(100)}{1}$$

Assuming you can approximate $C(q)$ by a differentiable function, then, since 1 is small relative to 100, the above quantity is approximately the same as $C'(100)$. Economists refer to $C'(q)$ as the *marginal cost*, the increase in cost incurred when you produce one more unit at a production level of q units.

Likewise, the *marginal revenue* $R'(q)$ gives the increase in revenue received when you sell one more unit at a sales level of q units.

In many instances, it is possible to formulate polynomial expressions that give $C(q)$ and $R(q)$ to an acceptable degree of accuracy. Suppose that, for the airline, the cost in dollars of operating q flights is

$$C(q) = 50,000 + 10,000q + 0.05q^2$$

The constant term \$50,000 represents the fixed overheads (rent, heat, some staffing costs, etc.). The linear term $\$10,000q$ is the primary cost per flight. The quadratic term $\$0.05q^2$ is the additional cost incurred due to the increased complexity of the operation.

The marginal cost function is

$$C'(q) = 10,000 + 0.1q$$

At $q = 100$,

$$C'(100) = 10,000 + 10 = \$10,010$$

Let's compare this with the actual cost increase given by computing the difference between $C(101)$ and $C(100)$:

$$C(101) - C(100) = 10,000(101) + 0.05(101^2) - 10,000(100) - 0.05(100^2)$$

This works out to be \$10,010.05. This is very close to the figure given by the marginal cost function. Thus, despite its simplicity, this example should be sufficient to illustrate why business planning is often done with polynomial (or other differentiable) approximations, using methods of calculus, where the manager can plot and examine graphs of $C(q)$, $R(q)$, $\pi(q)$, $C'(q)$, $R'(q)$, $\pi'(q)$, etc.

Let's see how to use such a mathematical model to minimize costs and to maximize profits.

We saw earlier that for the profit function $\pi(q)$ to be a maximum, we need

$$R'(q) = C'(q)$$

In words, for maximum profit, the production level should be such that the marginal revenue equals the marginal cost. To ensure that this is indeed a maximum profit, and not a minimum, we can use the second derivative test: For a maximum profit, the second derivative

$$\pi''(q) = R''(q) - C''(q)$$

should be negative. This requires

$$R''(q) < C''(q)$$

that is, the rate of increase of marginal revenue is less than the rate of increase of marginal cost.

The *average cost function* is defined as

$$ac(q) = \frac{C(q)}{q}$$

For the average cost to be a minimum, $ac'(q) = 0$. Differentiating using the quotient rule, we get

$$ac'(q) = \frac{qC'(q) - C(q)}{q^2} = 0$$

Thus, for a minimum average cost,

$$qC'(q) = C(q)$$

$$C'(q) = \frac{C(q)}{q} = ac(q)$$

Thus, for a minimum average cost, the marginal cost should equal the average cost.

4.2.3 Calculus in Human Anatomy

Calculus may be applied in the study of human anatomy. For example, because of friction at the walls, the flow of blood through a vein or an artery exhibits what is called *laminar flow*: the velocity of the blood is zero at the vessel wall and increases as you approach the central axis. In 1840, a French physician called Poiseuille discovered the following equation to describe the laminar flow of blood through a human blood vessel:

$$v = \frac{P}{4\eta L}(R^2 - r^2)$$

This equation assumes the blood vessel is a cylindrical tube of radius R and length L, where P is the pressure difference between the ends of the tube, η is the viscosity of the blood, r measures the distance from the central axis of the tube ($0 \leq r \leq R$), and v is the velocity of the blood at radius r.

The rate by which the velocity changes as you move outward from $r = 0$ to $r = R$ is called the *velocity gradient*. It is given by the equation

$$\text{velocity gradient } = \frac{dv}{dr} = -\frac{Pr}{2\eta L}$$

For a human artery, typical values of the various constants are: $\eta = 0.027$, $R = 0.008$ cm, $L = 2$ cm, $P = 4000$ dynes/cm^2. For these values,

$$v \approx 1.85 \times 10^4 \times (6.4 \times 10^{-5} - r^2)$$

At $r = 0.002$ cm, the blood is flowing with a velocity ($v(0.002)$) of about 1.11 cm/sec and the velocity gradient at that point is

$$-\frac{4000(0.002)}{2(0.027)2} \approx -74 \text{ cm/sec/cm}$$

4.2.4 Biological Growth

Table 4.1 gives the estimated population of Mexico for the period 1980–1986 (all population figures given in millions).

If you divide the population in any year by the population the previous year, the answer works out to be approximately 1.026. For instance,

$$\frac{\text{Population in 1982}}{\text{Population in 1981}} = \frac{70.93}{69.13} = 1.026$$

This means that the population growth is exponential; t years after 1980, the population (in millions) is given by the formula

$$P = 67.38(1.026)^t$$

If we assume that this growth pattern continues, the population will double every 27 years. In the year 2007 it will be 134.76 million, in 2034 it will be 269.52 million, and in 2061 it will be 539.04 million. We say that the population of Mexico has a *doubling time* of 27 years.

TABLE 4.1 Population of Mexico, 1980–1986

Year	Population	Change
1980	67.38	
1981	69.13	1.75
1982	70.93	1.80
1983	72.77	1.84
1984	74.66	1.89
1985	76.60	1.94
1986	78.59	1.99

Regular doubling of this kind is a feature of exponential growth. Let's take a look at it mathematically.

The *doubling time* for a population growing exponentially is the time it takes the population to double. If $P(t)$ grows according to the rule

$$P(t) = P_0 e^{rt}$$

(where P_0 is the initial population and r is a constant), the doubling time T is such that, for any fixed time t,

$$P_0 e^{r(t+T)} = 2P_0 e^{rt}$$

See Figure 4.7. We'll solve the above equation for T. In so doing, we'll verify that T does not depend on the choice of the start-time t; the same T works for any t.

$$\begin{aligned} e^{r(t+T)} &= 2e^{rt} \\ e^{rt} e^{rT} &= 2e^{rt} \\ e^{rT} &= 2 \\ rT &= \ln 2 \\ T &= \frac{\ln 2}{r} \end{aligned}$$

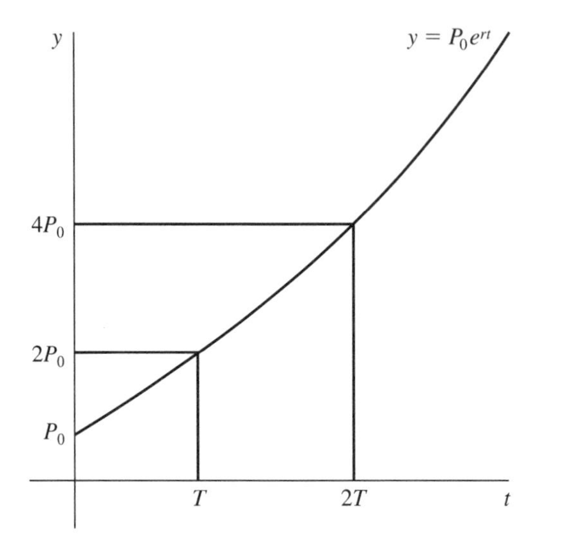

FIGURE 4.7 *Doubling time for a population* $P(t)$

4.2.5 Exponential Decay

Just as a population or substance that grows exponentially will have a fixed doubling time, so too a population or substance that decays exponentially will have a fixed half life, the time in which the size of the population or substance halves.

If the decay is described by the equation

$$M(t) = M_0 e^{-kt}$$

where M_0 is the initial population/mass and k is a positive constant, then to compute the *half life* T, we have to solve the equation

$$M_0 e^{-k(t+T)} = \tfrac{1}{2} M_0 e^{-kt}$$

for T, where t is any fixed number. (As we will see, the answer does not depend on t.) See Figure 4.8.

$$\begin{aligned} e^{-k(t+T)} &= \tfrac{1}{2} e^{-kt} \\ e^{-kt} e^{-kT} &= \tfrac{1}{2} e^{-kt} \\ e^{-kT} &= \tfrac{1}{2} \\ e^{kT} &= 2 \\ kT &= \ln 2 \\ T &= \frac{\ln 2}{k} \end{aligned}$$

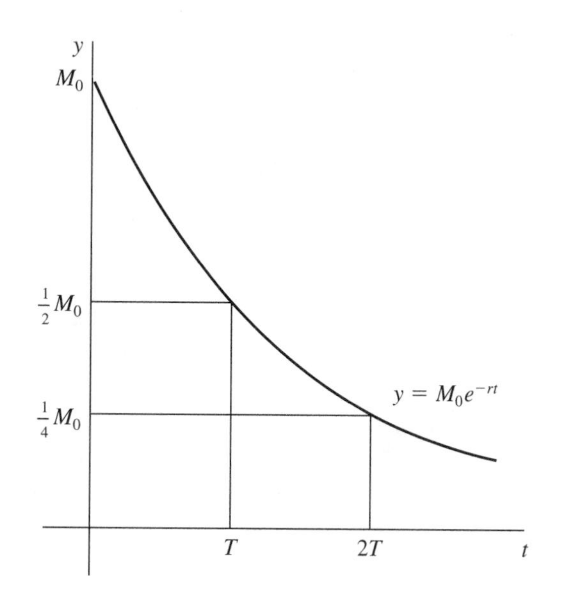

FIGURE 4.8 *Half-life of a mass* $M(t)$

For example, the half life of radium-226 is 1590 years. Suppose we start with a sample of radium-226 of mass 100 mg. Let's find a formula that tells us the mass remaining after t years. The formula will have the general form

$$M(t) = 100e^{-kt}$$

Since $M(1590) = 50$, we have

$$50 = 100e^{-1590k}$$

so

$$e^{1590k} = 2$$

This equation can be solved to give

$$k = \frac{\ln 2}{1590}$$

Thus, the mass of radium left after t years is given by the formula

$$M(t) = 100e^{-(\ln 2)t/1590}$$

For example, after 1,000 years, the mass remaining will be

$$M(1000) = 100e^{-(\ln 2)1000/1590} \approx 65 \text{ mg}.$$

Suppose we wanted to know how long it would be before the radium mass was reduced to 10 mg. Then we would have to solve the following equation for t:

$$10 = 100e^{-(\ln 2)t/1590}$$

Rewrite this equation as

$$e^{(\ln 2)t/1590} = 10$$

Taking logarithms,

$$\frac{(\ln 2)t}{1590} = \ln 10$$

Hence

$$t = \frac{\ln 10}{\ln 2} \times 1590 = \frac{2.30259}{0.69315} \times 1590 \approx 5282$$

years.

4.2.6 Problems

1. The relationship between distance (in meters) and time (in seconds) of a moving body is found to be approximated by the cubic equation

$$s = t^3 - 5t^2 + 8.5t$$

Find formulas for the velocity and acceleration of the body.

When is the body at rest? When is its speed a maximum or a minimum? Graph the distance, velocity, and acceleration functions, and use them to describe the motion.

2. A company has cost and price functions

$$C(q) = 3500 + 5q - 0.001q^2$$
$$p(q) = 50 - 0.01q$$

What production level will maximize profits?

3. The population of New Desert City is growing at 5% per year. In 1990, the population was 100,000. Assuming the same population growth continues, what will be the population in the year 2001? What will it be in 2050?

4. Polonium-210 has a half life of 140 days. If you start with a sample of 300 mg, what will the mass be after 100 days? How long will it take before the mass has decayed to 20 mg?

5. Concrete-walled basements of new houses often show a level of radioactivity due to the presence of radon-222. Simple measuring instruments are available to measure the level of radioactivity. Suppose that you use such a device, and you discover that the amount of radon-222 decays by 58% over a three-day period. What is the half life of radon-222, and how long will it take for the radon-222 to decay to 10% of its original level?

4.3 RELATED RATES

Sometimes, when we need to compute the rate of change of a quantity A, the best approach is to find an equation that relates A to a second quantity B, whose rate of change is either known or easy to compute. We then differentiate both sides of the equation with respect to time, and solve the new equation to obtain $A'(t)$ in terms of $B'(t)$. In this way we obtain $A'(t)$. Problems that can be solved in this fashion are called related rates problems.

4.3.1 Examples

Example 1

Helium is being pumped into a spherical weather balloon at a rate of 500 cc/sec. At what rate is the radius increasing when the diameter of the balloon is 100 cm?

Solution

The volume of the balloon is given by

$$V = \tfrac{4}{3}\pi r^3$$

We know the rate at which V is increasing:

$$\frac{dV}{dt} = 500$$

We want to find the value of $\frac{dr}{dt}$ when $r = 50$. Using the chain rule to differentiate the above formula for V,

$$\frac{dV}{dt} = \frac{4}{3}\pi\, 3r^2 \frac{dr}{dt} = 4\pi r^2 \frac{dr}{dt}$$

Hence,

$$\frac{dr}{dt} = \frac{1}{4\pi r^2}\frac{dV}{dt} = \frac{500}{4\pi r^2}$$

Putting $r = 50$, we get

$$\frac{dr}{dt} = \frac{500}{4\pi 2500} = \frac{1}{20\pi} \text{ cm/sec.}$$

Converting to cm/minute, when the diameter of the balloon is 100 cm, its radius is increasing at a rate of approximately $3/\pi$ or 0.955 cm/minute.

✎ *Example 2*

A street lamp at the top of a 20-ft pole shines down on a 6-ft man walking away from the pole at a constant speed of 5 ft/sec. How fast is the tip of the man's shadow moving when he is 30 ft from the pole?

Solution

Let m denote the distance of the man from the lamp post and let s denote the distance of the tip of his shadow from the lamp post. The situation is illustrated in Figure 4.9.

By similar triangles,

$$\frac{s}{20} = \frac{s - m}{6}$$

which can be solved to give

$$s = \frac{10}{7}m$$

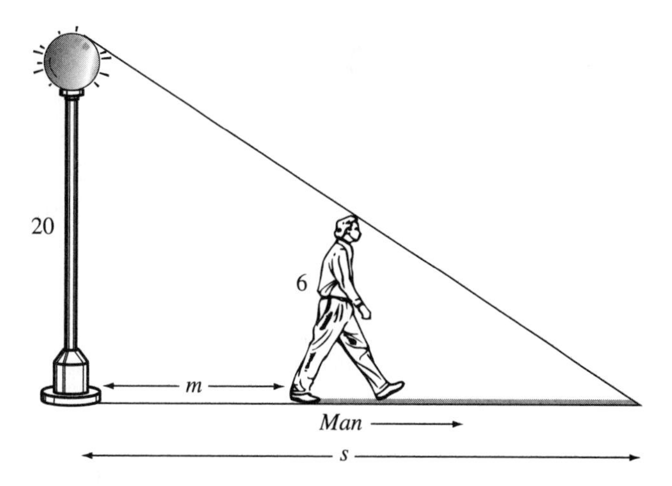

FIGURE 4.9 *The man, the lamp, and the shadow*

Differentiating this equation, we get

$$\frac{ds}{dt} = \frac{10}{7}\frac{dm}{dt}$$

But we know that $\frac{dm}{dt} = 5$ ft/sec. Hence

$$\frac{ds}{dt} = \frac{10}{7}5 = 7\tfrac{1}{7} \text{ ft/sec.}$$

This does not depend on m. At any instant, the tip of the shadow is moving away from the lamp at a speed of $7\frac{1}{7}$ ft/sec.

4.3.2 Problems

1. Alice is driving west along a straight road at 50 mph. Bob is driving north along a second straight road at 65 mph. Ahead of the two cars, the two roads intersect at a right angle. When Alice is $1/3$ mile from the junction, Bob is $1/2$ mile away. At what rate are the cars approaching each other at that moment?

2. A boat is being pulled into a jetty by a rope attached to the bow of the boat and passing through a pulley on the jetty that is 1 meter higher than the bow of the boat. The rope is being winched in at a rate of 1.5 meters per second. How fast is the boat approaching the jetty when it is 2 meters out?

4.4 IMPLICIT DIFFERENTIATION

Sometimes we are given y as a function of x *implicitly*, that is, in the form of an equation

$$f(x, y) = 0$$

involving x and y. For example, the equation of an ellipse, centered at the origin, may be given in the form

$$ax^2 + by^2 - 1 = 0$$

One way to find $\frac{dy}{dx}$ would be to first solve the equation to give y explicitly in terms of x. In some cases, this may be difficult, or even impossible.

An alternative approach is to differentiate the equation with respect to x and then solve for $\frac{dy}{dx}$. This alternative process is called *implicit differentiation*.

For example, differentiating the above equation of the ellipse we get

$$2ax + 2by\frac{dy}{dx} = 0$$

so, solving for $\frac{dy}{dx}$,

$$\frac{dy}{dx} = -\frac{ax}{by}$$

4.4.1 Examples

Example 1

Use the method of implicit differentiation to prove that for any positive rational number r,

$$\frac{d}{dx}(x^r) = rx^{r-1} \ (x > 0).$$

As we saw in the previous section, related rates problems often require implicit differentiation. Assume the result holds for positive integers r (as we proved in Topic 3, Section 3.1.3, using the binomial theorem). [Of course, we have proved the result for any real number r in Examples 3.4.1, Example 2—but that proof used deep properties of e^x and $\ln x$.

Solution

Let $y = x^{p/q}$, where p, q are positive integers. Then, by the properties of exponentiation,

$$y^q = x^p$$

Differentiating both sides of this equation with respect to x, using the chain rule on the left-hand side and the result for integers (which we have assumed) on the right,

$$qy^{q-1}\frac{dy}{dx} = px^{p-1}$$

Solving for $\frac{dy}{dx}$,

$$\frac{dy}{dx} = \frac{p}{q}\frac{x^{p-1}}{y^{q-1}}$$

$$= \frac{p}{q}\frac{x^{p-1}}{x^{p(q-1)/q}}$$

(Substituting for y in terms of x in the denominator.)

$$= \frac{p}{q}x^{(p-1)-(pq-p)/q}$$

(Combining numerator and denominator into a single power of x.)

$$= \frac{p}{q}x^{(pq-q-pq+p)/q}$$

(Expressing the exponent on x as a single quotient.)

$$= \frac{p}{q}x^{(p-q)/q}$$

(Simplifying the exponent on x.)

$$= \frac{p}{q}x^{(p/q)-1}$$

(Reformulating the exponent on x to give the desired form.)

as required.

✎ ***Example 2***

Use the method of implicit differentiation to prove that for any negative rational number r,

$$\frac{d}{dx}(x^r) = rx^{r-1}$$

Assume the result holds for positive rationals r (which we have just proved).

Solution

Suppose $y = x^{-r}$, where r is a positive rational number. Then $y = \frac{1}{x^r}$, and so $x^r y = 1$. Differentiating this last equation implicitly,

$$x^r \frac{dy}{dx} + rx^{r-1} y = 0$$

Substituting for y in terms of x,

$$x^r \frac{dy}{dx} + rx^{r-1} \frac{1}{x^r} = 0$$

Simplifying,

$$x^r \frac{dy}{dx} + rx^{-1} = 0$$

Dividing through by x^r,

$$\frac{dy}{dx} + rx^{-r-1} = 0$$

Solving for $\frac{dy}{dx}$,

$$\frac{dy}{dx} = -rx^{-r-1}$$

and the proof is complete.

4.4.2 Problems

1. Use implicit differentiation to find $\frac{dy}{dx}$ in each of the following cases:

(a) $x^3 + y^3 = 4xy$

(b) $\cos(x - y) = x^2 \sin y$

(c) $x\sqrt{1+y} + y\sqrt{1+x} = 2x$

4.5 DIFFERENTIALS

In Chapter 3, we introduced the terms dx and dy solely as constituents of the notation $\frac{dy}{dx}$ for the derivative; dx and dy were not given any independent meaning. In particular, $\frac{dy}{dx}$ was not regarded as a quotient of two numbers.

It is, however, possible to give dx and dy independent meaning, and this can be useful on occasion. The idea is to regard dx and dy as two new variables. Regarded in this way, dx and dy are called *differentials*.

4.5.1 Definitions

Given a differentiable function $y = f(x)$, take dx to be a completely new independent variable and dy to be a completely new dependent variable, related to dx by the equation

$$dy = f'(x)dx$$

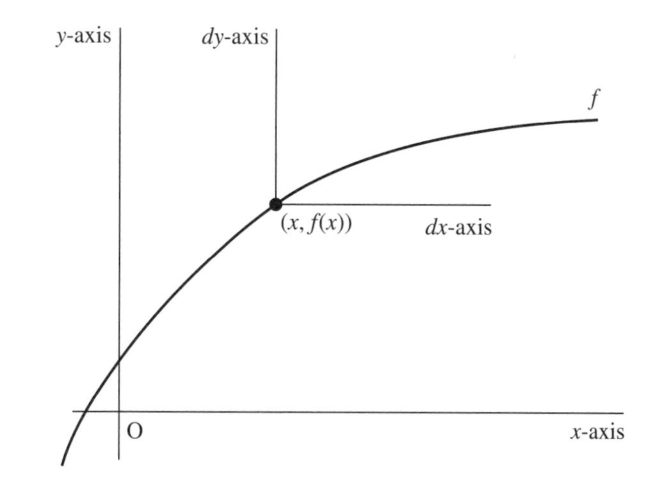

FIGURE 4.10 *Differential axes*

In this equation, we regard x as fixed; dx and dy are the variables.

In terms of the graph of f, we can think of dx and dy as providing a new pair of axes having their origin at the point $(x, f(x))$ on the curve (still regarding x as fixed). See Figure 4.10.

In terms of differentials, the expression

$$\frac{dy}{dx} = f'(x)$$

is not just a notational convenience. It is a genuine equation that specifies the value of a quotient of two quantities. (Remember, we regard x as a fixed number, so $f'(x)$ will be a number.)

Relative to the differential axes dx and dy, the equation

$$dy = f'(x)dx$$

determines a straight line that passes through the origin—that is, through the point $(x, f(x))$—which provides a straight-line approximation to the function f near the argument x. See Figure 4.11 on page 114.

4.5.2 Linear Approximation

We can use differentials to find approximate values of a differentiable function f near an argument a for which $f(a)$ and $f'(a)$ are known.

The idea is to take differentials at the point $(a, f(a))$ to give the following linear function approximating f:

$$dy = f'(a)dx$$

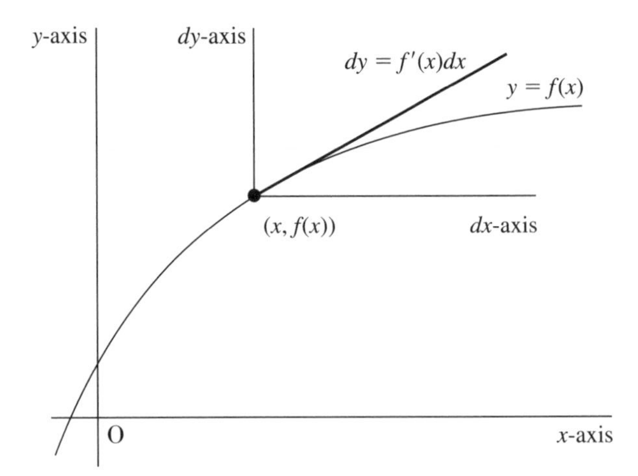

FIGURE 4.11 *Approximation by the tangent line*

Since the origin of the dx, dy axes is the point $(a, f(a))$, this gives the following approximation to $f(a+dx)$:

$$f(a+dx) \approx f(a) + f'(a)dx$$

Writing x instead of a, we get the general form of the formula:

$$f(x+dx) \approx f(x) + f'(x)dx$$

Referring to Figure 4.12 on page 115, the aim is to approximate the value of f at the point P, when we know the value $(f(a))$. The coordinates of P are $(a + dx, f(a + dx)$. For small x, the point Q is close to P. The coordinates of Q are $(a + dx, f(a) + dy)$, where $dy = f'(a)dx$. We take $f(a)+dy$ as our approximation for $f(a+dx)$.

4.5.3 Examples

Example 1

For the function

$$y = f(x) = 5x^3 - 3x^2 + 2$$

find dy when $x = 2$, $dx = 0.1$ and compare the result with the difference

$$f(2.1) - f(2)$$

Solution

Differentiating $f(x)$,

$$f'(x) = 15x^2 - 6x$$

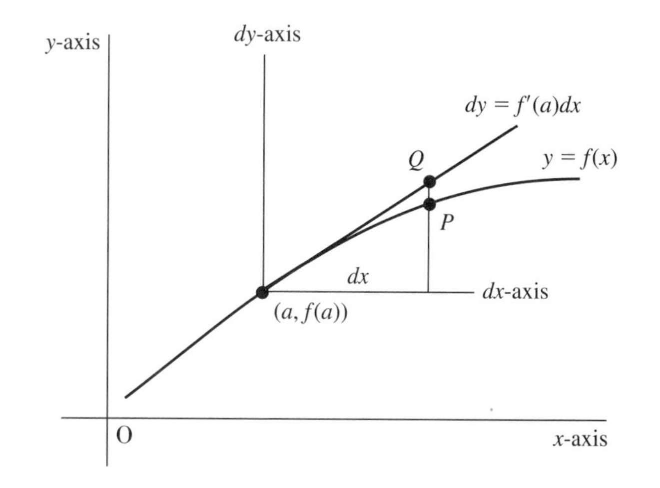

FIGURE 4.12 *Linear approximation by differentials*

At $x = 2$, $f'(2) = 48$, so the differentials satisfy the equation

$$dy = 48\,dx$$

For $dx = 0.1$,

$$dy = 4.8$$

Compare this with

$$f(2.1) - f(2) = 35.075 - 30 = 5.075$$

✎ ***Example 2*** Use differentials to find an approximate value of $\sqrt{36.1}$

Solution

We use the function

$$f(x) = \sqrt{x} = x^{1/2}$$

Differentiating,

$$f'(x) = \frac{1}{2}x^{-1/2} = \frac{1}{2\sqrt{x}}$$

Thus, at $x = 36$,

$$dy = \frac{1}{2\sqrt{36}}dx = \frac{1}{12}dx$$

But,

$$\sqrt{36.1} = f(36.1) = f(36 + 0.1)$$

Hence, taking $dx = 0.1$,

$$\sqrt{36.1} \approx f(36) + \frac{1}{12}0.1 = 6 + \frac{0.1}{12} \approx 6.00833$$

Squaring the result to check the accuracy:

$$6.00833^2 = 36.100029$$

which is correct to four decimal places.

4.5.4 Problems

1. Find the equation connecting the differentials and compute dy in each of the following cases:

   (a) $y = \sqrt{1-x},\ x = 0,\ dx = 0.01$

   (b) $y = \sin x,\ x = \pi/6,\ dx = 0.5$

   (c) $y = \dfrac{x-1}{x+1},\ x = 1,\ dx = -0.1$

2. Use differentials to approximate the following quantities:

   $$\sqrt[3]{218}, \quad \sqrt{99}, \quad \sin(59^\circ)$$

   (For the last one, express 59° in radians.)

4.6 PARAMETRIC CURVES

Sometimes we are given the coordinates (x, y) of a point P on a curve as functions $x = f(u),\ y = g(u)$ of a third variable (or *parameter*) u. For example, the equations

$$x = r\cos u, \quad y = r\sin u$$

determine a circle of radius r, with its center at the origin. In such a case, we say the equations

$$x = f(u), \quad y = g(u)$$

are *parametric equations* of the curve.

4.6.1 Method

If a curve is given in terms of a pair of parametric equations

$$x = f(u), \quad y = g(u)$$

the derivative dy/dx is given by:

$$\frac{dy}{dx} = \frac{dy}{du} \bigg/ \frac{dx}{du}$$

For example,

$$x = \cos u, \quad y = \sin^2 u$$

are parametric equations of the parabola $x^2 + y = 1$. Using the above formula,

$$\frac{dy}{dx} = \frac{2 \sin u \cos u}{-\sin u} = -2 \cos u = -2x$$

Of course, for this simple example, it is easy to obtain $\frac{dy}{dx}$ directly from the equivalent equation

$$y = 1 - x^2.$$

4.6.2 Problems

1. Find $\frac{dy}{dx}$ in terms of the given parameter in each of the following cases:

   (a) $x = 2(u - \sin u), \quad y = (1 - \cos u)$

   (b) $x = \sec u, \quad y = \tan u, \quad \text{for } -\frac{\pi}{2} < u < \frac{\pi}{2}$

   (c) $x = \frac{1 - t^2}{1 + t^2}, \quad y = \frac{2t}{1 + t^2}$

2. For each of the functions given in Problem 1, try to eliminate the parameter u to obtain a Cartesian equation of the curve.

3. Graph each of the curves in Problem 1.

4.7 INVERSE TRIG FUNCTIONS

Because they are periodic, the trigonometric functions do not have inverse functions. For every trigonometric function and for any argument A for which the function has a value, say r, the function gives the same value r for (at least) $A + 2\pi$, $A + 4\pi$, $A + 6\pi$, etc.

Nevertheless, there are some instances where it is useful to have a standard definition of an inverse function for a particular part of a trigonometric function, defined on a subdomain. For each number r in its domain, the inverse should pick out exactly one of the infinitely many possible angles (numbers) A sent to r by the original trigonometric function.

4.7.1 Definitions

The idea is to take each of the standard trig functions and restrict its domain to an interval on which the function is one-to-one (i.e., different arguments give different values), and yet on which interval every value of the function is achieved. In other words, we pick out an entire period of the function.

The inverse to the sine function is written

$$\arcsin X \text{ or } \sin^{-1} X$$

It is defined on the domain $[-1, 1]$ with range $[-\pi/2, \pi/2]$. The definition is

$$y = \arcsin x \Leftrightarrow \sin y = x$$

The inverse to the cosine function is written

$$\arccos X \text{ or } \cos^{-1} X$$

It is defined on the domain $[-1, 1]$ with range $[0, \pi]$. The definition is

$$y = \arccos x \Leftrightarrow \cos y = x$$

The inverse to the tangent function, $\arctan X$ or $\tan^{-1} X$, is defined on the domain $(-\infty, \infty)$ with range $(-\pi/2, \pi/2)$. The definition is

$$y = \arctan x \Leftrightarrow \tan y = x$$

Graphs of all three functions are given in Figure 4.13.

There are similar definitions of arcsecant, arccosecant, and arccotangent. However, for our purposes, we only require the ones above.

4.7.2 Differentiation

We calculate the derivatives of the inverse trig functions by using the inverse function rule.

Theorem. $\dfrac{d}{dx} \arcsin x = \dfrac{1}{\sqrt{1 - x^2}}$

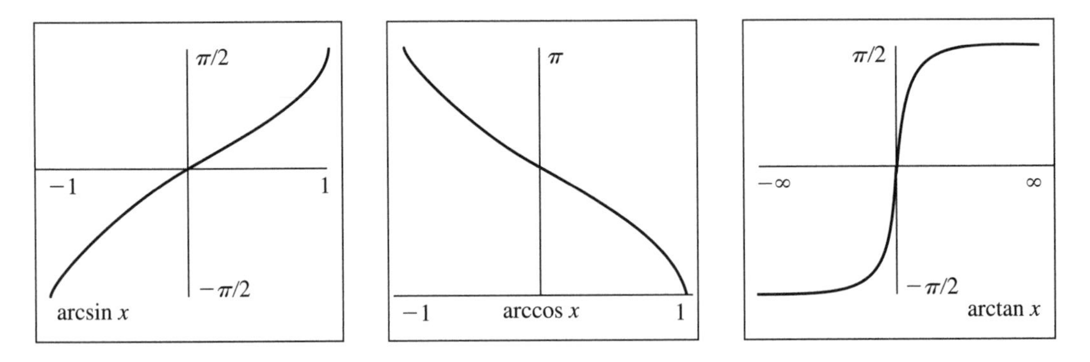

FIGURE 4.13 *Inverse trigonometric functions*

Proof

Let $y = \arcsin x$. Then $x = \sin y$, so $\dfrac{dx}{dy} = \cos y$. By the inverse function rule,

$$\frac{dy}{dx} = 1\Big/\frac{dx}{dy} = \frac{1}{\cos y} = \frac{1}{\sqrt{1-\sin^2 y}} = \frac{1}{\sqrt{1-x^2}}$$

We can write $\cos y = \sqrt{1-\sin^2 y}$ (i.e., taking the positive square root) because the choice of the domain of $\arcsin x$ makes $\arcsin x$ increasing on its domain. □

Theorem.

$$\frac{d}{dx}\arccos x = -\frac{1}{\sqrt{1-x^2}}$$

Proof

Let $y = \arccos x$. Then $x = \cos y$, so $\dfrac{dx}{dy} = -\sin y$. By the inverse function rule,

$$\frac{dy}{dx} = 1\Big/\frac{dx}{dy} = -\frac{1}{\sin y} = -\frac{1}{\sqrt{1-\cos^2 y}} = -\frac{1}{\sqrt{1-x^2}}$$

We can write $\sin y = \sqrt{1-\cos^2 y}$ (i.e., taking the positive square root), because the choice of the domain of $\arccos x$ makes $\arccos x$ decreasing on its domain. □

Theorem.

$$\frac{d}{dx}\arctan x = \frac{1}{1+x^2}$$

Proof

Let $y = \arctan x$. Then $x = \tan y$, so $\dfrac{dx}{dy} = \sec^2 y$. By the inverse function rule,

$$\frac{dy}{dx} = 1\Big/\frac{dx}{dy} = \frac{1}{\sec^2 y} = \frac{1}{1+\tan^2 y} = \frac{1}{1+x^2}$$

The proof is complete. □

4.7.3 Problems

1. Find $f'(x)$ in each of the following cases. Simplify your answer as much as possible.

(a) $f(x) = \arcsin(2x)$

(b) $f(x) = \arcsin^2 x$

(c) $f(x) = (\arccos x)(\ln x)$

(d) $f(x) = \dfrac{\arctan x}{1 + x^2}$

(e) $f(x) = \arccos\left(\dfrac{1}{x}\right)$

dx

topic 5

Integration I

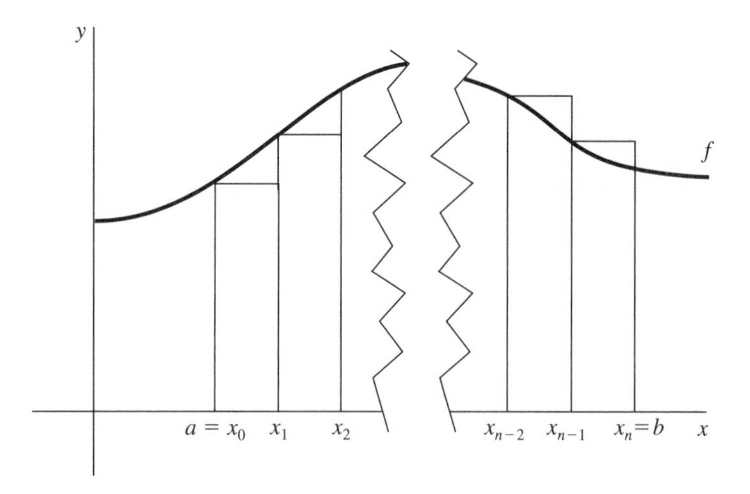

FIGURE 5.1 *Definition of the definite integral*

5.1 THE DEFINITE INTEGRAL

Integral calculus is a collection of procedures that (among other uses) enable us to calculate the areas and volumes of various geometric figures. The idea is to regard the figure as made up of an infinite number of infinitely small constituents, for each of which we know how to calculate the area or volume. What makes this idea work is its close connection with differentiation.

Among the many other uses of integral calculus are computing the amount of oil pumped through a pipeline at a variable rate, calculating the work done in moving an object, determining the center of mass of a planar object, and calculating the average profit of a business over a period of time.

5.1.1 Computing Areas

Suppose we want to compute the area A bounded by the curve $y = f(x)$, the x-axis, and the lines $x = a$, $x = b$. A first step is to divide the interval $[a, b]$ into n equally wide subintervals

$$[x_0, x_1], \ldots, [x_{n-1}, x_n]$$

where we set $x_0 = a$, $x_n = b$ for notational convenience. The length of each subinterval is the length of the entire interval divided by the number of subintervals, that is,

$$\frac{b - a}{n}$$

For each $i = 0, \ldots, n - 1$, draw the vertical line from x_i to the curve and complete the rectangle to the right of the vertical (Figure 5.1). The height of the ith rectangle is $f(x_i)$, the value of the function at x_i.

Thus the area of the ith rectangle (width times height) is

$$\frac{b-a}{n} \times f(x_i)$$

Hence the total area of all n rectangles is

$$A(n) = \sum_{i=0}^{n-1} \frac{b-a}{n} f(x_i)$$

As the number n of subintervals increases, and the length

$$\frac{b-a}{n}$$

of each subinterval decreases, the area

$$A(n) = \sum_{i=0}^{n-1} \frac{b-a}{n} f(x_i)$$

(i.e., the total area of all the rectangles) approaches the area A we want to compute. Taking the limit, we obtain an exact expression for the area A:

$$A = \lim_{n\to\infty} \sum_{i=0}^{n-1} \frac{b-a}{n} f(x_i)$$

5.1.2 Definition of the Definite Integral

In general, regardless of whether we are setting out to compute an area, the quantity

$$\lim_{n\to\infty} \sum_{i=0}^{n-1} \frac{b-a}{n} f(x_i)$$

(if this limit exists) is called the *definite integral* of the function $f(x)$ from a to b. It is usually denoted by

$$\int_a^b f(x)\,dx$$

The numbers a and b are called the *limits of integration.*

Note that $\int_a^b f(x)\,dx$ is a *number*, not a function.

5.1.3 Example

✎ ***Example 1***

Let's use the definition of the integral to compute

$$\int_0^1 x^2 dx$$

[This example is not typical. It was chosen specially because we can actually work out the exact answer in this case.]

Solution

By the definition,

$$\int_0^1 x^2\,dx = \lim_{n\to\infty} A(n)$$

where

$$A(n) = \frac{1}{n}\left[0^2 + \left(\frac{1}{n}\right)^2 + \left(\frac{2}{n}\right)^2 + \left(\frac{3}{n}\right)^2 + \cdots + \left(\frac{n-1}{n}\right)^2\right]$$

[You should check to see that this expression for $A(n)$ is the same as

$$\sum_{i=0}^{n-1} \frac{b-a}{n} f(x_i)$$

in this case.] The above expression for $A(n)$ simplifies to:

$$A(n) = \frac{1}{n^3}\left[0^2 + 1^2 + 2^2 + 3^2 + \cdots + (n-1)^2\right]$$

There is a standard formula for the sum of the squares of the first $n-1$ natural numbers:

$$1^2 + 2^2 + 3^2 + \cdots + (n-1)^2 = \frac{1}{6}(n-1)n(2n-1)$$

Using this formula in the expression for $A(n)$, we get

$$A(n) = \frac{1}{6n^3}(n-1)n(2n-1)$$

Hence,

$$\lim_{n\to\infty} A(n) = \lim_{n\to\infty} \frac{1}{6}\left(1 - \frac{1}{n}\right)\left(2 - \frac{1}{n}\right) = \frac{1}{3}$$

5.1.4 Problems

1. Use a computer or a programmable calculator to estimate the value of

$$\int_1^2 x^2\,dx$$

by the following method, based on the definition of the definite integral. For $n = 5, 10, 50, 100, 500, 1000$, calculate the quantity

$$A(n) = \frac{1}{n}\left[1^2 + \left(1 + \frac{1}{n}\right)^2 + \left(1 + \frac{2}{n}\right)^2 + \cdots + \left(1 + \frac{n-1}{n}\right)^2\right]$$

Then estimate the limit of the sequence of values $A(5)$, $A(10)$, $A(50)$, $A(100)$, $A(500)$, $A(1000)$, ...

2. Use the same method as in Problem 1 to estimate the value of

$$\int_0^2 x^3\,dx$$

3. Now do the same for

$$\int_0^1 e^{x^2}\,dx$$

(For those who have read ahead, note that this one can't be done by finding an antiderivative!)

5.2 THE FUNDAMENTAL THEOREM OF CALCULUS

Using the definition

$$\int_a^b f(x)dx = \lim_{n\to\infty} \sum_{i=0}^{n-1} \frac{b-a}{n} f(x_i)$$

of the integral to compute (say) an actual area would require the computation of numerous (large) finite sums and the estimation of their limit. This would be extremely time consuming. However, for some functions f there is an alternative approach. It turns out that there is a close connection between integration and differentiation. That connection is what is called the Fundamental Theorem of Calculus.

Theorem (The Fundamental Theorem of Calculus). Let f be a continuous function, defined on the interval $[a, b]$. Then

$$\int_a^b f(x)dx = F(b) - F(a)$$

where $F(x)$ is any function such that $F'(x) = f(x)$. ■

Given $f(x)$, a function $F(x)$ such that $F'(x) = f(x)$ is sometimes called an *antiderivative* of $f(x)$. By the fundamental theorem of calculus, one way to evaluate a definite integral of $f(x)$ is to find an antiderivative of $f(x)$.

5.2.1 Proof of the Fundamental Theorem

For definiteness, we shall prove the theorem for a function $f(x)$ that is positive on the interval $[a, b]$. The general case is entirely similar.

For any positive real number x between a and b, let $A(x)$ denote the area bounded by the curve, the x-axis, the line $x = a$, and the vertical line at x. See Figure 5.2(1). This defines a function $A(x)$.

Notice that

$$A(b) = \int_a^b f(x)dx$$

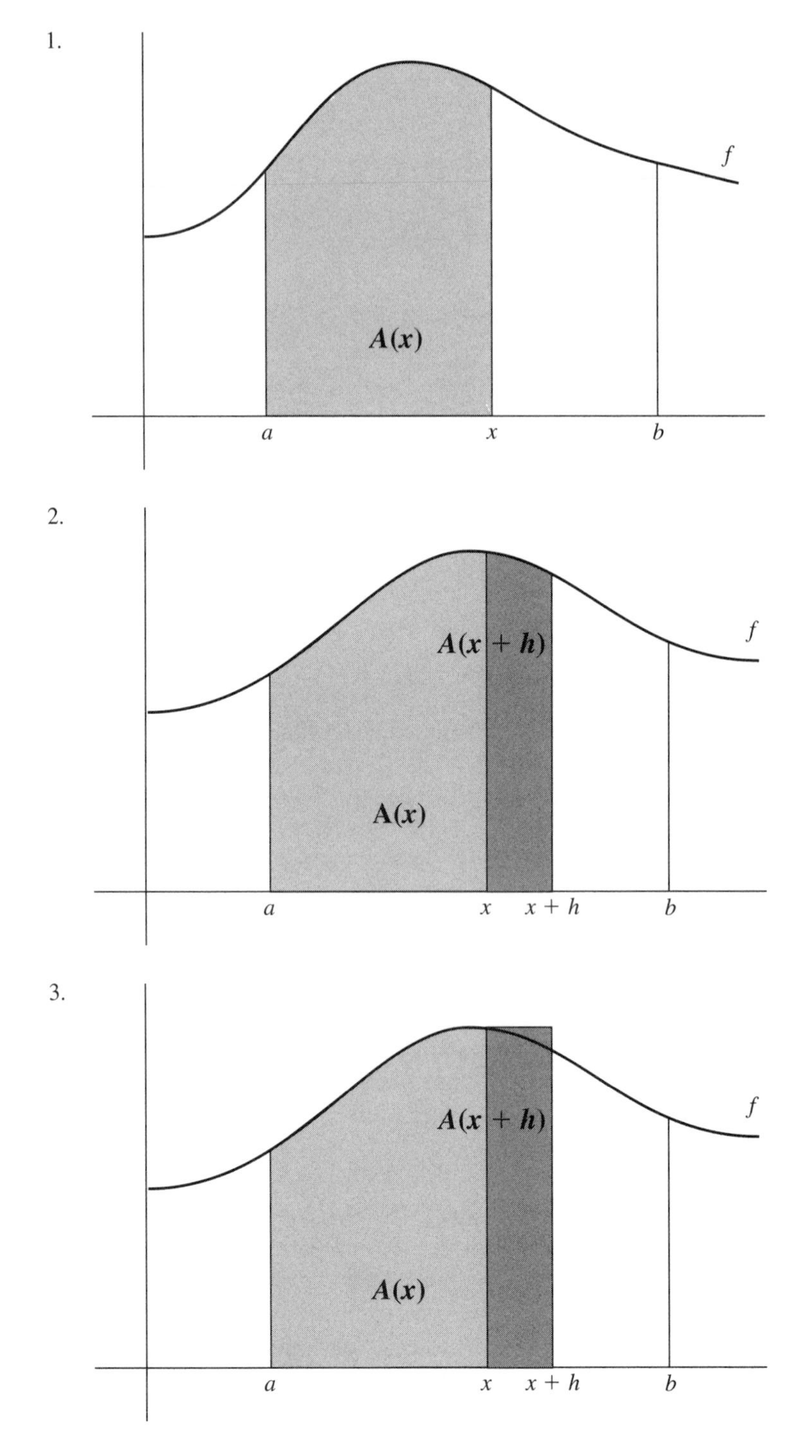

FIGURE 5.2 *Proof of the fundamental theorem of calculus*

Consider the change in the value of $A(x)$ when x is increased by a small amount h. The darker shaded region in Figure 5.2(2) is the difference between $A(x)$ and $A(x+h)$.

If h is small, the difference between $A(x)$ and $A(x+h)$ is approximately equal to the area of the rectangle of width h and height $f(x)$ (see Figure 5.2(3)), namely $hf(x)$. That is:

$$A(x+h) - A(x) \approx hf(x)$$

where $\approx$ denotes approximate equality

Dividing by h,

$$\frac{A(x+h) - A(x)}{h} \approx f(x)$$

The smaller we make h, the smaller becomes the error in the above approximation. In the limit, we have

$$\lim_{h \to 0} \frac{A(x+h) - A(x)}{h} = f(x)$$

But the limit expression on the left is just the definition of the derivative $A'(x)$ of the function $A(x)$. Hence, what we have shown is that:

$$A'(x) = f(x)$$

In other words, $A(x)$ is an antiderivative of $f(x)$.

Suppose now that $F(x)$ is any antiderivative of $f(x)$. Thus

$$F'(x) = f(x) = A'(x)$$

Thus

$$[F - A]'(x) = F'(x) - A'(x) = 0$$

This means that the function $F - A$ is constant on $[a, b]$, say

$$[F - A](x) = C$$

So,

$$F(x) = A(x) + C$$

Since $A(a) = 0$, we can determine the constant C by setting $x = a$:

$$F(a) = 0 + C$$

Thus

$$F(x) = A(x) + F(a)$$

It follows that

$$\int_a^b f(x)\,dx = A(b) = F(b) - F(a)$$

And the proof is complete. □

5.2.2 Examples

✎ ***Example 1***

Use the fundamental theorem of calculus to evaluate

$$\int_3^5 x^2\,dx$$

Solution

Start by observing that for the function

$$F(x) = \frac{x^3}{3}$$

we have

$$F'(x) = x^2$$

Thus $F(x)$ is an antiderivative of x^2. Hence, by the fundamental theorem of calculus,

$$\begin{aligned}\int_3^5 x^2\,dx &= F(5) - F(3)\\ &= \frac{125}{3} - \frac{27}{3}\\ &= \frac{98}{3} = 32\tfrac{2}{3}\end{aligned}$$

✎ ***Example 2***

Evaluate

$$\int_0^{\pi/4} \sin x\,dx$$

Solution

The function $F(x) = -\cos x$ is an antiderivative of $\sin x$. Hence,

$$\begin{aligned}\int_0^{\pi/4} \sin x\,dx &= F(\tfrac{\pi}{4}) - F(0)\\ &= (-\cos(\tfrac{\pi}{4})) - (-\cos 0)\\ &= (-\tfrac{1}{\sqrt{2}}) - (-1)\\ &= 1 - \tfrac{1}{\sqrt{2}}\end{aligned}$$

5.2.3 Problems

1. Use the fundamental theorem of calculus to evaluate the following integrals:

   (a) $\int_0^2 x^3\, dx$

   (b) $\int_0^{\pi/2} \cos x\, dx$

   (c) $\int_0^1 e^x\, dx$

   (d) $\int_0^{10} x\, dx$

   (e) $\int_1^e \frac{1}{x}\, dx$

5.3 INDEFINITE INTEGRALS

By virtue of the fundamental theorem of calculus, given a continuous function $f(x)$, a function $F(x)$ such that $F'(x) = f(x)$ is often called an *indefinite integral* of $f(x)$. Thus, the terms 'indefinite integral' and 'antiderivative' are synonymous. The indefinite integral of $f(x)$ is denoted by

$$\int f(x)dx$$

In fact, there is not just one indefinite integral of $f(x)$, but a family; if $F(x)$ is any antiderivative of $f(x)$, then any function of the form

$$F(x) + C$$

is also an antiderivative of $f(x)$, where C is any constant.

In general, whenever we evaluate an indefinite integral, we should include an unknown constant, generally referred to as a *constant of integration.* Occasionally, we have additional information that enables us to determine the exact value of this constant; but in general, (indefinite) integration produces a function that includes an unknown constant.

5.3.1 Basic Rules for Integration

The integration rules we describe in this section are all immediate consequences of the corresponding rules for differentiation.

- $\int x^n dx = \frac{x^{n+1}}{n+1} + C$, if $n \neq -1$, where C is a constant of integration.
- $\int [f(x) + g(x)]dx = \int f(x)dx + \int g(x)dx$
- $\int af(x)dx = a \int f(x)dx$, for any constant a.

A special case of the first rule is that

$$\int 1 dx = x + C$$

So, using the third rule, for any constant a,

$$\int a\, dx = ax + C$$

Using the above rules, we can integrate any polynomial function. Thus, for example:

$$\int [3x^7 - 25x^4 - 11x + 5]dx = \frac{3}{8}x^8 - 5x^5 - \frac{11}{2}x^2 + 5x + C$$

- $\int \frac{1}{x} dx = \ln|x| + C$
- $\int e^x dx = e^x + C$

More generally, $\int e^{kx} dx = \frac{e^{kx}}{k} + C$

- $\int a^x dx = \frac{a^x}{\ln a} + C, \qquad a > 0, \quad a \neq 1$

More generally, $\int a^{kx} dx = \frac{a^{kx}}{k \ln a} + C, \qquad a > 0, \quad a \neq 1$

- $\int \sin x\, dx = -\cos x + C$

More generally, $\int \sin kx\, dx = -\frac{1}{k} \cos kx + C$

- $\int \cos x\, dx = \sin x + C$

More generally, $\int \cos kx\, dx = \frac{1}{k} \sin kx + C$

5.3.2 Examples

Example 1

Evaluate

$$\int \frac{x^3+1}{\sqrt{x}}dx$$

Solution

The idea is to rewrite the integral in terms of powers of x.

$$\begin{aligned}\int \frac{x^3+1}{\sqrt{x}}dx &= \int \frac{x^3+1}{x^{1/2}}dx \\ &= \int [x^{5/2}+x^{-1/2}]\,dx \\ &= \frac{x^{7/2}}{7/2}+\frac{x^{1/2}}{1/2}+C \\ &= \tfrac{2}{7}x^{7/2}+2x^{1/2}+C \\ &= \tfrac{2}{7}\sqrt{x}(x^3+7)+C\end{aligned}$$

Example 2

Evaluate

$$\int [7\sin 3x+5\cos 2x+e^{5x}+9]\,dx$$

Solution

It's hard to imagine this particular integral ever arising in practice, but it does include a lot of the basic cases. Here is the answer:

$$-\tfrac{7}{3}\cos 3x+\tfrac{5}{2}\sin 2x+\tfrac{1}{5}e^{5x}+9x+C$$

Example 3

Evaluate

$$\int \left[x^3+x^2+x+1+\frac{1}{x}+\frac{1}{x^2}+\frac{1}{x^3}\right]dx$$

Solution

The idea is to rewrite the integral in terms of powers of x (apart from $\dfrac{1}{x}$):

$$\int \left[x^3+x^2+x+1+\frac{1}{x}+x^{-2}+x^{-3}\right]dx$$

This is evaluated as:

$$\frac{x^4}{4} + \frac{x^3}{3} + \frac{x^2}{2} + x + \ln|x| + \frac{x^{-1}}{-1} + \frac{x^{-2}}{-2} + C$$

which tidies up to give:

$$\frac{x^4}{4} + \frac{x^3}{3} + \frac{x^2}{2} + x + \ln|x| - \frac{1}{x} - \frac{1}{2x^2} + C$$

5.3.3 Problems

1. Evaluate the following integrals:

(a) $\int [x^{11} - 3x^8 + 21x^2 - 5]dx$ **(b)** $\int 7^x\,dx$

(c) $\int [11\sin 3x - 7\cos 5x]dx$ **(d)** $\int \sin \pi x\,dx$

(e) $\int e\,dx$ **(f)** $\int e^{8.2x}dx$

(g) $\int \frac{x^3+1}{x^2}dx$ **(h)** $\int \sqrt{x}(1-x)dx$

(i) $\int \frac{3}{x}dx$ **(j)** $\int \frac{1}{x^5}dx$

5.4 TRIGONOMETRIC INTEGRALS

Integrals involving trig functions can sometimes be evaluated using the following trigonometric identities:

$$\sin^2 x = \tfrac{1}{2}(1-\cos 2x) \qquad \cos^2 x = \tfrac{1}{2}(1+\cos 2x)$$

$$\sin x\cos x = \tfrac{1}{2}\sin 2x$$

$$\sec^2 x = 1 + \tan^2 x \qquad \csc^2 x = 1 + \cot^2 x$$

For example,

$$\int \cos^2 x\,dx = \int \tfrac{1}{2}(1+\cos 2x)\,dx$$
$$= \tfrac{1}{2}x + \tfrac{1}{4}\sin 2x + C$$

Again,

$$\int \sin x \cos x \, dx = \int \tfrac{1}{2} \sin 2x \, dx$$
$$= \tfrac{1}{4} \cos 2x + C$$

5.4.1 Examples

✎ ***Example 1***

Evaluate the integral

$$\int \tan^2 x \, dx$$

Solution

We use the identity $\sec^2 x = 1 + \tan^2 x$ to transform the integral as follows:

$$\int \tan^2 x \, dx = \int [\sec^2 x - 1] \, dx$$
$$= \tan x - x + C$$

✎ ***Example 2***

Evaluate $\int \cos^3 x \, dx$

Solution

The first step is to transform the integral like this:

$$\int \cos^3 x \, dx = \int \cos^2 x \cos x \, dx$$

Magic? No, not really. But it is rather clever. Let's finish the solution and then see what is going on. We rewrite the right-hand side to give

$$\int \cos^3 x \, dx = \int (1 - \sin^2 x) \cos x \, dx$$
$$= \int \cos x \, dx - \int \sin^2 x \cos x \, dx$$
$$= \sin x \quad (+ \text{ a constant}) - \int \sin^2 x \cos x \, dx$$

Now we can use the fact that, by the chain rule

$$\frac{d}{dx}(\sin^3 x) = (3 \sin^2 x)(\cos x)$$

from which it follows that

$$\int \sin^2 x \cos x \, dx = \tfrac{1}{3} \sin^3 x \ \ (+ \text{ a constant})$$

Hence,

$$\int \cos^3 x \, dx = \sin x - \tfrac{1}{3} \sin^3 x + C$$

The idea behind this solution is to manipulate the original integral into a form that we can recognize as a derivative using the chain rule. To do this, you need to be familiar with the basic derivatives. Even so, it still takes a flash of inspiration. Integration can take time, and sometimes you need to try several different approaches until you find one that works. As with most other things in life, the more experience you have, the better and quicker you get. But integration is not like arithmetic, or like differentiation, for that matter. You can generally solve arithmetic or differentiation problems by applying the rules in a systematic fashion. As an inverse process (namely, *anti*differentiation), integration can sometimes require a bit of ingenuity. Sometimes nothing works.

5.4.2 Problems

1. Evaluate the following integrals:

(a) $\displaystyle\int \sin^2 x \, dx$

(b) $\displaystyle\int \sin^3 x \, dx$

(c) $\displaystyle\int \sin^2 x \cos^2 x \, dx$

(d) $\displaystyle\int \sin^4 x \, dx$

(e) $\displaystyle\int \sin^5 x \, dx$

5.5 EVALUATING DEFINITE INTEGRALS

By the fundamental theorem of calculus, to compute a definite integral

$$\int_a^b f(x) dx$$

you first compute the indefinite integral (i.e., find an antiderivative) $F(x)$ of the function $f(x)$, using the techniques of integral calculus. Then you set

$$\int_a^b f(x)dx = F(b) - F(a)$$

The 'bracket notation' is often used in this connection:

$$[F(x)]_a^b = F(b) - F(a)$$

For instance,

$$\left[x^3\right]_2^4 = 4^3 - 2^3 = 64 - 8 = 56$$

Note that there is no need to include a constant of integration when computing a definite integral. To see why, let's look at a simple example.

We evaluate the integral

$$\int_1^4 x^2 dx$$

The first step is to integrate (i.e., antidifferentiate) x^2 to obtain the indefinite integral

$$F(x) = \frac{x^3}{3}$$

Notice that we have not included a constant of integration. Then, using the bracket notation,

$$\int_1^3 x^2 dx = F(4) - F(1) = \left[\frac{x^3}{3}\right]_1^4 = \frac{64}{3} - \frac{1}{3} = \frac{63}{3} = 21$$

Now let's see why we were able to ignore the constant of integration. Suppose we did include a constant of integration, writing

$$F(x) = \frac{x^3}{3} + C$$

Then, when we evaluate the difference $F(2) - F(0)$, we get

$$F(2) - F(0) = \left[\frac{64}{3} + C\right] - \left[\frac{1}{3} + C\right] = \frac{63}{3} = 21$$

This is the same answer as before. The two instances of C cancel out in the subtraction. This will always be the case. The evaluation of a definite integral from an indefinite integral always involves a subtraction of two values, so there is never any need to include a constant of integration. But be careful. You can only ignore the constant of integration when you are evaluating definite integrals using the fundamental theorem of calculus. On other occasions when you calculate an indefinite integral, the constant of integration can be critical, and should not be omitted.

5.5.1 Problems

1. Evaluate the following definite integrals:

 (a) $\displaystyle\int_{-1}^{1} [x^3 - 4x^2 + 3x - 1]\,dx$

 (b) $\displaystyle\int_{-\pi}^{\pi} \sin^2 x\,dx$

 (c) $\displaystyle\int_{0}^{4} e^{2x}\,dx$

 (d) $\displaystyle\int_{0}^{2} \sqrt{x}\,dx$

 (e) $\displaystyle\int_{1}^{3} \frac{1}{x}\,dx$

5.6 INTEGRATION BY SUBSTITUTION

Consider the integral

$$\int 4x^3 \cos(x^4)dx$$

This has the property that the function outside the cosine function, namely $4x^3$, is the derivative of the argument of the cosine function, x^4. The integral may be evaluated by making the substitution

$$u = x^4 \text{ (so, in terms of differentials, } du = 4x^3 dx)$$

which transforms the integral into the much simpler form

$$\int \cos u\,du$$

This is a special (and very simple) example of the *method of (algebraic) substitution.*

5.6.1 Algebraic Substitution

Algebraic substitution works when we are faced with an integral of the form

$$\int f(h(x))h'(x)dx$$

consisting of a function f applied to another function h, multiplied by the derivative h' of h. The method is to simplify the integral by making the substitution

$$u = h(x), \quad du = h'(x)dx$$

Integration by substitution is a consequence of the chain rule for differentiation. Here is the theorem the method depends upon.

Theorem. If $f(x)$ is a continuous function and $h(x)$ is a differentiable function, and if we set

$$u = h(x)$$

then

$$\int f(h(x))h'(x)dx = \int f(u)du$$

Proof

Let F be an antiderivative of f. Then $F' = f$.

Let $u = h(x), \ y = F(u)$. Thus $y = F(h(x))$.

By the chain rule,

$$\begin{aligned} \frac{dy}{dx} &= \frac{dy}{du}\frac{du}{dx} = F'(u)h'(x) \\ &= f(u)h'(x) = f(h(x))h'(x) \end{aligned}$$

Integrating this equation with respect to x,

$$y = \int f(h(x))h'(x)dx$$

But F is an antiderivative of f, so

$$y = F(u) = \int f(u)du$$

Thus, equating these two expressions for y,

$$\int f(h(x))h'(x)dx = \int f(u)du \qquad \square$$

Not all integrals for which substitution can be used have exactly the form of the above theorem. You sometimes have to do some preliminary work. For example, to evaluate the integral

$$\int \frac{x^3}{x^2 - 1} dx$$

we first rewrite it as

$$\tfrac{1}{2} \int \frac{x^2}{x^2 - 1} 2x \, dx$$

and then try the substitution $u = x^2 - 1$. Then, $du = 2x\,dx$, so the integral becomes

$$\begin{aligned}\tfrac{1}{2}\int \frac{u+1}{u}\,du &= \tfrac{1}{2}\int \left[1+\frac{1}{u}\right]du \\ &= \tfrac{1}{2}[u + \ln|u|\,] + C \\ &= \tfrac{1}{2}[x^2 - 1 + \ln|x^2-1|\,] + C \\ &= \tfrac{1}{2}x^2 - \tfrac{1}{2} + \tfrac{1}{2}\ln|x^2-1| + C\end{aligned}$$

Since C is an arbitrary constant, we may as well incorporate the lone $\frac{1}{2}$ into it, to give the answer as

$$\tfrac{1}{2}x^2 + \tfrac{1}{2}\ln|x^2-1| + C$$

We may use the method of substitution to evaluate $\int \tan x\,dx$. On the face of it, this does not look like a candidate for substitution, until you write it as

$$\int \frac{\sin x}{\cos x}\,dx = \int \frac{1}{\cos x}\sin x\,dx$$

We make the substitution $u = \cos x$. Then, $du = -\sin x\,dx$. Hence,

$$\begin{aligned}\int \tan x\,dx &= -\int \frac{1}{u}du \\ &= -\ln|u| + C \\ &= -\ln|\cos x| + C\end{aligned}$$

Thus,

- $\int \tan x\,dx = -\ln|\cos x| + C$

More generally, we have

$$\int \tan kx\,dx = -\frac{1}{k}\ln|\cos kx| + C$$

In a similar fashion, we can also evaluate the following integral:

- $\int \cot x\,dx = \ln|\sin x| + C$

With the above two results, we have extended our list of integrals of the basic trigonometric functions. The only two basic trig functions we have not yet integrated are $\sec x$ and $\csc x$. Both are handled by means of an ingenious trick. For $\sec x$, we have:

$$\int \sec x\,dx = \int \frac{\sec x(\sec x + \tan x)}{\sec x + \tan x}dx$$

TABLE 5.1 Trigonometric Substitutions	
Integral Contains	**Substitute**
$\sqrt{a^2 - x^2}$	$x = a \sin z$
$\sqrt{a^2 + x^2}$	$x = a \tan z$
$\sqrt{x^2 - a^2}$	$x = a \sec z$

$$= \int \frac{\sec x \tan x + \sec^2 x}{\sec x + \tan x} dx$$

Now we make the substitution $u = \sec x + \tan x$. For this substitution, we have $du = (\sec x \tan x + \sec^2 x)dx$. Thus the integral becomes

$$\int \frac{du}{u} = \ln |u| + C = \ln |\sec x + \tan x| + C$$

Thus,

- $\displaystyle\int \sec x \, dx = \ln |\sec x + \tan x| + C$

  Similarly,

- $\displaystyle\int \csc x \, dx = \ln |\csc x - \cot x| + C$

5.6.2 Trigonometric Substitutions

Integrals that involve any one of the expressions

$$\sqrt{a^2 - x^2}$$
$$\sqrt{x^2 - a^2}$$
$$\sqrt{a^2 + x^2} \text{ (or } \sqrt{x^2 + a^2}\text{)}$$

can sometimes be evaluated by making a trigonometric substitution, which transforms the integral from an algebraic one to one involving trigonometric functions.

Table 5.1 gives the appropriate substitutions to try. After the integral has been evaluated, you need to substitute x back for z. You can often use the diagrams in Figure 5.3 to make this substitution.

For example, to evaluate

$$\int \frac{dx}{x\sqrt{4 - x^2}}$$

we try the substitution $x = 2 \sin z$. For this substitution, we have

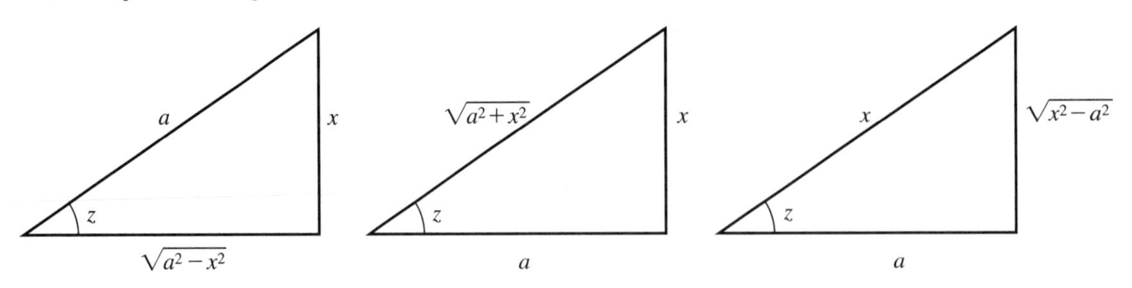

FIGURE 5.3 *Substitution diagrams*

$dx = 2\cos z\,dz$, and so

$$\begin{aligned}\int \frac{dx}{x\sqrt{4-x^2}} &= \int \frac{2\cos z\,dz}{2\sin z\sqrt{4-4\sin^2 z}} \\ &= \int \frac{2\cos z\,dz}{2\sin z\sqrt{4(1-\sin^2 z)}} \\ &= \int \frac{2\cos z\,dz}{2\sin z\sqrt{4\cos^2 z}} \\ &= \int \frac{2\cos z\,dz}{2\sin z\,2\cos z} \\ &= \int \frac{dz}{2\sin z} \\ &= \frac{1}{2}\int \csc z\,dz\end{aligned}$$

But the above integral is that of a standard basic trigonometric function that we saw how to integrate at the end of the last section. It works out to be

$$\frac{1}{2}\ln|\csc z - \cot z| + C$$

Referring to Figure 5.4, this transforms into

$$\frac{1}{2}\ln\left|\frac{2}{x} - \frac{\sqrt{4-x^2}}{x}\right| + C = \frac{1}{2}\ln\left|\frac{2-\sqrt{4-x^2}}{x}\right| + C$$

and we are done.

Incidentally, substitution of $\sin z$ for x will also let you evaluate

$$\int \frac{dx}{\sqrt{1-x^2}}$$

but in this case we have already seen that

$$\frac{d}{dx}\arcsin x = \frac{1}{\sqrt{1-x^2}}$$

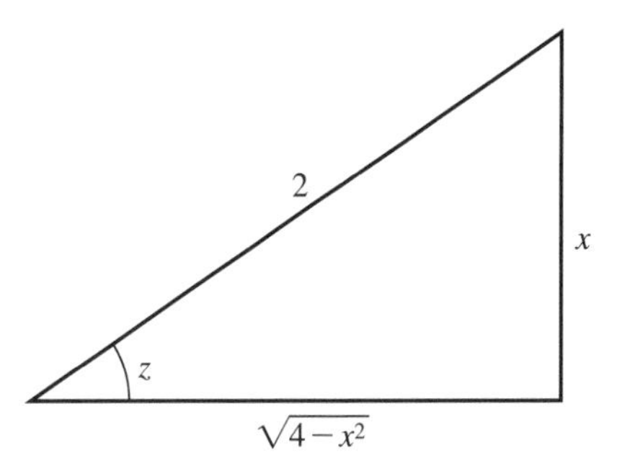

FIGURE 5.4 *Substitution diagram for* $x = 2\sin z$

so we can conclude immediately that

$$\int \frac{dx}{\sqrt{1-x^2}} = \arcsin x + C$$

Moral: It's always worth making sure you are not overlooking the "obvious."

5.6.3 Substitution and Limits of Integration

Suppose we want to evaluate the definite integral

$$\int_a^b f(x)\,dx$$

Suppose further that in order to evaluate the indefinite integral

$$\int f(x)\,dx$$

we make the substitution $u = h(x)$. This transforms the integral into the form

$$\int G(u)\,du$$

We can then evaluate this integral, substitute back for x, and then plug in the limits of integration a and b.

But sometimes there is a better method. We can use the substitution formula $u = h(x)$ to transform the limits of integration. If we set

$$c = h(a), d = h(b)$$

the definite integral becomes

$$I = \int_c^d G(u)du$$

For example, to evaluate

$$\int_0^2 x^3 \cos(x^4) dx$$

we make the substitution $u = x^4$. For this substitution, $du = 4x^3\, dx$, so $x^3\, dx = \frac{1}{4} du$. Moreover,

$$x = 0 \Rightarrow u = 0$$
$$x = 2 \Rightarrow u = 16$$

So the integral transforms to

$$\begin{aligned} \int_0^{16} \cos u\, du &= [\sin u]_0^{16} \\ &= \sin 16 - \sin 0 \\ &= \sin 16 \end{aligned}$$

5.6.4 Examples

✎ ***Example 1***

Evaluate the integral

$$\int \frac{x}{x^2+1}\, dx$$

Solution

We make the substitution $u = x^2+1$. For this substitution, $du = 2x\, dx$, so $x\, dx = \frac{1}{2} du$. The integral transforms to:

$$\frac{1}{2}\int \frac{1}{u}\, du = \frac{1}{2}\ln|u| + C = \frac{1}{2}\ln|x^2+1| + C$$

✎ ***Example 2***

Evaluate the integral

$$\int x^2\sqrt{x^3+1}\, dx$$

Solution

We make the substitution $u = x^3+1$. For this substitution, $du = 3x^2\, dx$, so $x^2\, dx = \frac{1}{3} du$. The integral transforms to:

$$\frac{1}{3}\int u^{1/2}\, du = \frac{1}{3}\frac{u^{3/2}}{3/2} + C = \frac{2}{9}(x^3+1)^{3/2} + C$$

Example 3

Evaluate the integral

$$\int_0^1 x\sqrt{1-x^2}\,dx$$

Solution

We make the substitution $u = 1 - x^2$. For this substitution, $du = -2x\,dx$, so $x\,dx = -\frac{1}{2}du$. Moreover,

$$x = 0 \Rightarrow u = 1$$
$$x = 1 \Rightarrow u = 0$$

So, the integral transforms to:

$$-\frac{1}{2}\int_1^0 u^{1/2}\,du = -\frac{1}{2}\left[\frac{u^{3/2}}{3/2}\right]_1^0 = -\frac{1}{2}\left[\frac{2}{3}u^{3/2}\right]_1^0 = -\frac{1}{2}\left[0-\frac{2}{3}\right] = \frac{1}{3}$$

Example 4

Evaluate the integral

$$\int_0^1 \frac{dx}{1+x}$$

Solution

We make the substitution $u = 1 + x$. For this substitution, $du = dx$. Moreover,

$$x = 0 \Rightarrow u = 1$$
$$x = 1 \Rightarrow u = 2$$

So, the integral transforms to:

$$s\int_1^2 \frac{du}{u} = [\ln|u|\,]_1^2 = \ln 2 - \ln 1 = \ln 2$$

Example 5

Evaluate the integral

$$\int_0^{\sqrt{\pi}} x\sin(x^2)\,dx$$

Solution

We make the substitution $u = x^2$. For this substitution, $du = 2x\,dx$, so $x\,dx = \frac{1}{2}du$. Moreover,

$$x = 0 \Rightarrow u = 0$$
$$x = \sqrt{\pi} \Rightarrow u = \pi$$

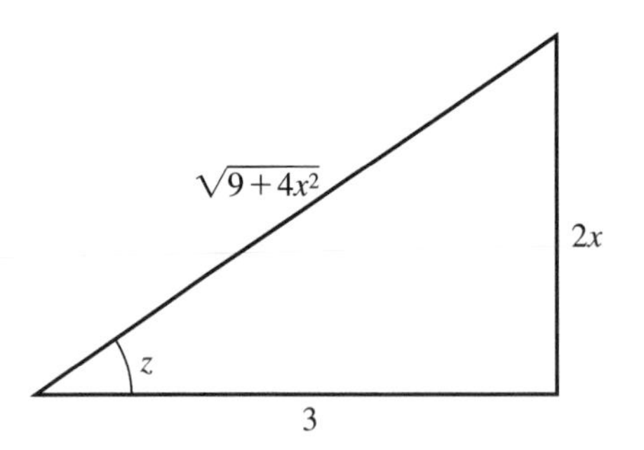

FIGURE 5.5 *Substitution diagram for Example 6*

So, the integral transforms to:

$$\tfrac{1}{2}\int_0^{\pi} \sin u\, du = \tfrac{1}{2}\big[-\cos u\big]_0^{\pi} = \tfrac{1}{2}[(1)-(-1)] = \tfrac{1}{2}[2] = 1$$

Example 6

Evaluate the integral

$$\int \frac{dx}{x\sqrt{9+4x^2}}$$

Solution

We make the substitution $x = \frac{3}{2}\tan z$. Then $dx = \frac{3}{2}\sec^2 z\, dz$ and $\sqrt{9+4x^2} = 3\sec z$. So:

$$\begin{aligned}\int \frac{dx}{x\sqrt{9+4x^2}} &= \int \frac{\frac{3}{2}\sec^2 z\, dz}{(\frac{3}{2}\tan z)(3\sec z)}\\ &= \tfrac{1}{3}\int \csc z\, dz\\ &= \tfrac{1}{3}\ln|\csc z - \cot z| + C\end{aligned}$$

Referring to Figure 5.5, $\csc z = \dfrac{\sqrt{9+4x^2}}{2x}$ and $\cot z = \dfrac{3}{2x}$, so the integral transforms to

$$\frac{1}{3}\ln\left|\frac{\sqrt{9+4x^2}}{2x} - \frac{3}{2x}\right| + C = \frac{1}{3}\ln\left|\frac{\sqrt{9+4x^2}-3}{2x}\right| + C$$

and we are done.

Example 7

Evaluate

$$\int \frac{dx}{1+x^2}$$

Solution

Remember the moral we saw earlier of making sure you don't overlook the "obvious." Here it is again. We already know that

$$\frac{d}{dx}\arctan x = \frac{1}{1+x^2}$$

Hence, without any further work, we have

$$\int \frac{dx}{1+x^2} = \arctan x + C$$

5.6.5 Problems

1. Evaluate the following integrals:

(a) $\int (x+1)^{12}\,dx$

(b) $\int \frac{x^2}{\sqrt{x^3+1}}\,dx$

(c) $\int \frac{e^{1/x}}{x^2}\,dx$

(d) $\int \frac{x^2}{\sqrt{x^2-25}}\,dx$

(e) $\int \frac{dx}{x^2\sqrt{3+x^2}}$

(f) $\int \frac{dx}{x\sqrt{4+9x^2}}$

(g) $\int \frac{\sqrt{9x^2+4}}{x}\,dx$

(h) $\int \frac{dx}{(1-x^2)^{3/2}}$

(i) $\int \cot x\,dx$

2. Evaluate the following integrals:

(a) $\int_1^2 x^2(x^3+4)^2\,dx$

(b) $\int_1^4 \frac{dx}{2x+1}$

(c) $\int_0^{0.5} \frac{x}{\sqrt{1-x^2}}\,dx$

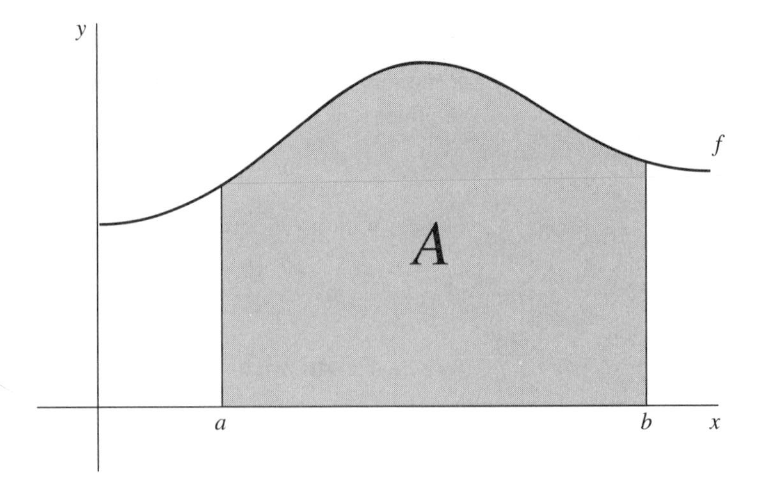

FIGURE 5.6 *Computing the area under a curve*

5.7 AREAS UNDER CURVES

Computing the area under a curve is one of the motivating applications of integration, and the one we used to develop the definition of the integral.

If a continuous function $f(x)$ is positive on the interval (a, b) (i.e., $a < x < b$), then the area bounded by the curve, the x-axis, and the lines $x = a$, $x = b$ (area A in Figure 5.6) is given by the definite integral

$$\int_a^b f(x)\,dx$$

If $f(x)$ is negative on $a < x < b$, the area is given by

$$-\int_a^b f(x)\,dx$$

If $f(x)$ changes sign on (a, b), you first identify the points where the curve crosses the x-axis, say at $x = c$ and $x = d$ (as shown in Figure 5.7), and then compute the individual areas between a and c, between c and d, and between d and b. The required area is the sum of these individual areas.

5.7.1 Examples

✎ ***Example 1***

Compute the area bounded by the curve

$$y = x^3 - 3x^2 + 2x$$

and the interval $[0, 2]$ of the x-axis.

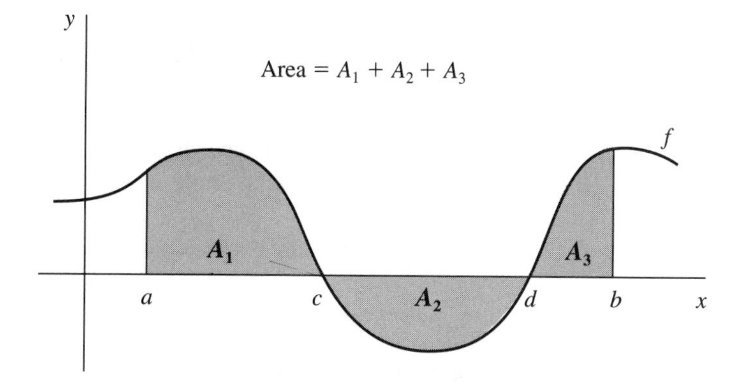

FIGURE 5.7 *Computing the area bounded by a curve*

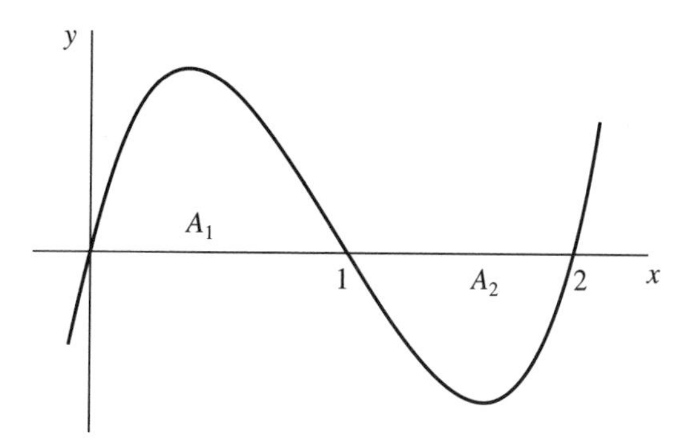

FIGURE 5.8 *Graph of* $y = x^3 - 3x^2 + 2x$

Solution

The curve crosses the x-axis at $x = 0, 1, 2$. See Figure 5.8. The indefinite integral of the function is

$$\int y\,dx = \int [x^3 - 3x^2 + 2x]\,dx = \frac{x^4}{4} - x^3 + x^2$$

Hence,

$$\begin{aligned}\text{Area } A_1 &= \left[\frac{x^4}{4} - x^3 + x^2\right]_0^1 \\ &= \left(\frac{1}{4} - 1 + 1\right) - (0) = \frac{1}{4}\end{aligned}$$

$$\text{Area } A_2 = -\left[\frac{x^4}{4} - x^3 + x^2\right]_1^2$$

$$= -\left[(4-8+4) - \left(\frac{1}{4} - 1 + 1\right)\right] = -\left[-\frac{1}{4}\right] = \frac{1}{4}$$

Thus, the total area is $\frac{1}{4} + \frac{1}{4} = \frac{1}{2}$

Notice that you have to evaluate two integrals to get this area. If you try to do it as a single integral, you get

$$\int_0^2 y\,dx = \left[\frac{x^4}{4} - x^3 + x^2\right]_0^2 = [(4-8+4) - (0)] = 0$$

which is wrong.

5.7.2 Problem

1. Evaluate the area bounded by the curve

$$y = x^3 - 5x^2 + 6x$$

and the interval $[0, 3]$ of the x-axis. [*Hint.* The function factors as $x(x-2)(x-3)$.]

dx

topic 6

Integration II

6.1 INTEGRATION BY PARTS

Integration by parts is a useful technique, in which the task of evaluating a given integral is replaced by the evaluation of one derivative and two new integrals. The idea is to do this so that the three new calculus problems are all significantly easier to handle than the given integral.

6.1.1 Method

Integration by parts depends on the following formula. If u and v are differentiable functions of x, then

$$\int u\,dv = uv - \int v\,du$$

This formula is derived from the product rule for differentiation:

$$\frac{d}{dx}[uv] = u\,\frac{dv}{dx} + v\,\frac{du}{dx} = u\,v' + v\,u'$$

Integrating the product rule identity,

$$uv = \int [u\,v']dx + \int [v\,u']dx$$

Now, $\frac{du}{dx}$ and u' are two notations for the same function (namely, the derivative of u), so we have the notational identity $\frac{du}{dx} = u'$. So let us agree to extend this notation by writing du for the expression $u'\,dx$. Similarly, we will write dv for the expression $v'\,dx$. Then the above integral equation can be rewritten as

$$uv = \int u\,dv + \int v\,du$$

Rearranging this identity gives

$$\int u\,dv = uv - \int v\,du$$

which is the formula for integration by parts.

The key to successful use of integration by parts is to split the integral up so that it becomes simpler. Since the first step is to differentiate u and integrate dv, you have to take care that neither of these leads to a more complicated integral than the original one.

Sometimes, you have to apply the parts formula two or more times in succession.

See Example 4 on p. 156.

6.1.2 Examples

✎ Example 1

Evaluate

$$\int xe^x\,dx$$

Solution

Let

$$u = x \qquad dv = e^x\,dx$$
$$du = dx \qquad v = e^x$$

Then,

$$\begin{aligned}\int xe^x\,dx &= \int u\,dv = uv - \int v\,du\\ &= xe^x - \int e^x\,dx\\ &= xe^x - e^x + C\\ &= e^x(x-1) + C\end{aligned}$$

Example 2

Evaluate

$$\int x \sin x\,dx$$

Solution

Let

$$u = x \qquad dv = \sin x\,dx$$
$$du = dx \qquad v = -\cos x$$

Then,

$$\begin{aligned}\int x \sin x\,dx &= \int u\,dv = uv - \int v\,du\\ &= -x\cos x - \int (-\cos x)\,dx\\ &= -x\cos x + \int \cos x\,dx\\ &= -x\cos x + \sin x + C\\ &= \sin x - x\cos x + C\end{aligned}$$

Example 3

Evaluate

$$\int x^2 \ln x\,dx$$

Solution

Let

$$u = \ln x \qquad dv = x^2\,dx$$
$$du = \frac{1}{x}\,dx \qquad v = \frac{x^3}{3}$$

Then,

$$\begin{aligned}
\int x^2 \ln x\,dx &= \int u\,dv = uv - \int v\,du \\
&= \frac{x^3}{3}\ln x - \int \left(\frac{x^3}{3}\right)\frac{1}{x}\,dx \\
&= \frac{x^3}{3}\ln x - \frac{1}{3}\int x^2\,dx \\
&= \frac{x^3}{3}\ln x - \frac{1}{3}\left(\frac{x^3}{3}\right) + C \\
&= \frac{x^3}{3}\ln x - \frac{x^3}{9} + C \\
&= \frac{x^3}{9}(3\ln x - 1) + C
\end{aligned}$$

✎ *Example 4*

Evaluate

$$\int x^2 \sin x\,dx$$

Solution

Let

$$u = x^2 \qquad dv = \sin x\,dx$$
$$du = 2x\,dx \qquad v = -\cos x$$

Then,

$$\begin{aligned}
\int x^2 \sin x\,dx &= \int u\,dv = uv - \int v\,du \\
&= -x^2\cos x - \int (-\cos x)\,2x\,dx \\
&= -x^2\cos x + 2\int x\cos x\,dx
\end{aligned}$$

Now we use integration by parts again to evaluate $\int x\cos x\,dx$. Let

$$u = x \qquad dv = \cos x\,dx$$
$$du = dx \qquad v = \sin x$$

Then,

$$\int x\cos x\,dx = \int u\,dv = uv - \int v\,du$$
$$= x\sin x - \int \sin x\,dx$$
$$= x\sin x + \cos x$$

(We won't worry about the constant of integration at this intermediate point.) Hence,

$$\int x^2 \sin x\,dx = -x^2\cos x + 2[x\sin x + \cos x] + C$$
$$= -x^2\cos x + 2x\sin x + 2\cos x + C$$

✎ ***Example 5***

Evaluate

$$\int \ln x\,dx$$

Solution

This doesn't look like an integral to which we can apply the method of integration by parts. But as it turns out, this is an easy way to evaluate this integral. Let

$$u = \ln x \qquad dv = dx$$
$$du = \frac{1}{x}\,dx \qquad v = x$$

Then,

$$\int \ln x\,dx = \int u\,dv = uv - \int v\,du$$
$$= x\ln x - \int x\,\frac{1}{x}\,dx$$
$$= x\ln x - \int dx$$
$$= x\ln x - x + C$$
$$= x(\ln x - 1) + C$$

✎ ***Example 6***

Evaluate

$$I = \int e^x \sin x\,dx$$

Solution

Let

$$u = e^x \qquad dv = \sin x\, dx$$

$$du = e^x\, dx \qquad v = -\cos x$$

Then

$$I = \int e^x \sin x\, dx = \int u\, dv = uv - \int v\, du$$

$$= -e^x \cos x + \int e^x \cos x\, dx$$

Now we have to evaluate the integral

$$J = \int e^x \cos x\, dx$$

This is very similar to the integral I we started with, and at this point the more fainthearted might be tempted to give up, thinking we have not made any progress. But if we press ahead, success is just around the next corner. Here goes.

To evaluate J, let

$$u = e^x \qquad dv = \cos x\, dx$$

$$du = e^x\, dx \qquad v = \sin x$$

Then,

$$J = \int e^x \cos x\, dx = \int u\, dv = uv - \int v\, du$$

$$= e^x \sin x - \int e^x \sin x\, dx$$

$$= e^x \sin x - I$$

Hence

$$I = -e^x \cos x + J = -e^x \cos x + e^x \sin x - I$$

This is a simple equation for I. We can rewrite it as

$$2I = e^x(\sin x - \cos x)$$

which gives

$$I + \tfrac{1}{2}e^x(\sin x - \cos x)$$

Adding the constant of integration, we get the general solution

$$I = \tfrac{1}{2}e^x(\sin x - \cos x) + C$$

Neat, huh?

6.1.3 Problems

1. Evaluate the following integrals. Some require a combination of the method of integration by parts with other methods of integration to evaluate the second integral $\int v\,du$.

   (a) $\int x\sqrt{1+x}\,dx$

   (b) $\int x^2 e^x\,dx$

   (c) $\int x^3 e^{2x}\,dx$

   (d) $\int \ln(x^2+1)\,dx$

   (e) $\int \sec^3 x\,dx$ (*Hint:* Write $\sec^3 x$ as $(\sec x)(\sec^2 x)$.)

2. Use the method of integration by parts to evaluate the following integrals.

   (a) $\int \arcsin x\,dx$

   (b) $\int \arccos x\,dx$

   (c) $\int \arctan x\,dx$

6.2 INTEGRATION BY PARTIAL FRACTIONS

A function obtained by dividing one polynomial by another is called a *rational* function. For example,

$$f(x) = \frac{x^3 - 1}{x^2 + 1}$$

A rational function may be integrated by the method of partial fractions. This involves splitting up the rational function into a sum of a polynomial and one or more 'partial fractions', rational functions of a particularly simple kind, each of which can be integrated separately.

6.2.1 Method

A rational function $\frac{p(x)}{q(x)}$, where $p(x)$ and $q(x)$ are polynomials, is said to be *proper* if the degree of $p(x)$ is less than the degree of $q(x)$.

Given a nonproper rational function you want to integrate, first divide the denominator into the numerator to obtain the sum of a polynomial and a proper rational function. For example,

$$\frac{x^3 - 1}{x^2 + 1} = x - \frac{x + 1}{x^2 + 1}$$

Then split up the rational part into a sum of *partial fractions*. These are rational functions with denominators of the form $(ax+b)^n$ or $(ax^2 + bx + c)^n$.

To split a proper rational function into partial fractions, first factor the denominator into a product of linear factors $ax+b$ and irreducible quadratic factors $ax^2 + bx + c$. (A quadratic is called *irreducible* if it cannot be factored into a product of two (real) linear polynomials.)

When the denominator has been factored, the proper rational function can be split up into a sum of partial fractions using the following four rules, depending on the nature of the denominator.

Rule 1. For each linear factor $ax + b$ that occurs just once in the denominator, there is a single partial fraction of the form

$$\frac{A}{ax + b}$$

where A is a constant that has to be determined.

Rule 2. For each linear factor $ax + b$ that occurs n times in the denominator, there is a sum of n partial fractions of the form

$$\frac{A_1}{ax + b} + \frac{A_2}{(ax + b)^2} + \cdots + \frac{A_n}{(ax + b)^n}$$

where the A_i are constants to be determined.

Rule 3. For each irreducible quadratic factor ax^2+bx+c that occurs just once in the denominator, there is a single partial fraction of the form

$$\frac{Ax + B}{ax^2 + bx + c}$$

where A, B are constants to be determined.

Rule 4. For each irreducible quadratic factor ax^2+bx+c that occurs n times in the denominator, there is a sum of n partial fractions of the form

$$\frac{A_1x + B_1}{ax^2 + bx + c} + \frac{A_2x + B_2}{(ax^2 + bx + c)^2} + \cdots + \frac{A_nx + B_n}{(ax^2 + bx + c)^n}$$

where the A_i, B_i are constants to be determined.

6.2.2 Examples

✎ ***Example 1***

Evaluate the integral

$$\int \frac{dx}{x^2 - 4}$$

Solution

First factor the denominator as

$$x^2 - 4 = (x - 2)(x + 2)$$

Using Rule 1 for partial fractions,

$$\frac{1}{x^2 - 4} = \frac{A}{x - 2} + \frac{B}{x + 2}$$

for constants A, B. Cross multiplying to clear the denominators,

$$1 = A(x + 2) + B(x - 2)$$

Simplifying,

$$1 = (A + B)x + 2(A - B)$$

Comparing coefficients of x,

$$0 = A + B$$

Comparing constant terms,

$$1 = 2A - 2B$$

Solving these simultaneous equations for A, B,

$$A = \tfrac{1}{4}, \ B = -\tfrac{1}{4}$$

So,

$$\frac{1}{x^2 - 4} = \frac{1}{4}\frac{1}{x - 2} - \frac{1}{4}\frac{1}{x + 2}$$

It follows that

$$\begin{aligned}\int \frac{dx}{x^2 - 4} &= \frac{1}{4}\int \frac{dx}{x - 2} - \frac{1}{4}\int \frac{dx}{x + 2} \\ &= \tfrac{1}{4}\ln|x - 2| - \tfrac{1}{4}\ln|x + 2| + C \\ &= \tfrac{1}{4}\ln\left|\frac{x - 2}{x + 2}\right| + C\end{aligned}$$

✎ *Example 2*

Evaluate the integral

$$\int \frac{x^2}{1-x^4}\,dx$$

Solution

Factor the denominator as

$$1-x^4=(1-x)(1+x)(1+x^2)$$

Using Rules 1 and 3 for partial fractions,

$$\frac{x^2}{1-x^4}=\frac{A}{1-x}+\frac{B}{1+x}+\frac{Cx+D}{1+x^2}$$

for constants A, B, C, D. Multiplying by

$$(1-x^4)=(1+x^2)(1+x)(1-x)$$

to clear the denominators,

$$x^2=A(1+x)(1+x^2)+B(1-x)(1+x^2)+(Cx+D)(1-x)(1+x)$$

Setting $x=1$,

$$1=4A, \quad \text{so} \quad A=\tfrac{1}{4}$$

Setting $x=-1$,

$$1=4B, \quad \text{so} \quad B=\tfrac{1}{4}$$

Setting $x=0$,

$$0=A+B+D, \quad \text{so} \quad D=-\tfrac{1}{2}$$

Equating coefficients of x,

$$0=A-B+C, \quad \text{so} \quad C=0$$

So,

$$\begin{aligned}\int \frac{x^2}{1-x^4}\,dx &= \frac{1}{4}\int\frac{dx}{1-x}+\frac{1}{4}\int\frac{dx}{1+x}-\frac{1}{2}\int\frac{dx}{1+x^2}\\ &= -\tfrac{1}{4}\ln|1-x|+\tfrac{1}{4}\ln|1+x|-\tfrac{1}{2}\arctan x+C\\ &= \tfrac{1}{4}\ln\left|\frac{1+x}{1-x}\right|-\tfrac{1}{2}\arctan x+C\end{aligned}$$

Notice how we chose values of x to make various terms vanish in order to determine the values of A, B, C, D. This is a useful trick to remember.

Example 3

Evaluate the integral

$$\int \frac{x^2 + 3x - 4}{x^2 - 2x - 8}\,dx$$

Solution

Write

$$\frac{x^2 + 3x - 4}{x^2 - 2x - 8} = 1 + \frac{5x + 4}{x^2 - 2x - 8}$$

Factor $x^2 - 2x - 8$ as

$$x^2 - 2x - 8 = (x + 2)(x - 4)$$

Using Rule 1 for partial fractions,

$$\frac{5x + 4}{x^2 - 2x - 8} = \frac{A}{x + 2} + \frac{B}{x - 4}$$

for constants A, B. Cross multiplying to clear the denominators,

$$5x + 4 = A(x - 4) + B(x + 2)$$

Setting $x = -2$,

$$-6 = -6A, \quad \text{so} \quad A = 1$$

Setting $x = 4$,

$$24 = 6B, \quad \text{so} \quad B = 4$$

So,

$$\begin{aligned}\int \frac{x^2 + 3x - 4}{x^2 - 2x - 8}\,dx &= \int 1\,dx + \int \frac{dx}{x + 2} + 4\int \frac{dx}{x - 4} \\ &= x + \ln|x + 2| + 4\ln|x - 4| + C \\ &= x + \ln|x + 2|\,|x - 4|^4 + C\end{aligned}$$

Example 4

Evaluate the integral

$$\int \frac{2x + 1}{x^3 - x^2 - x + 1}\,dx$$

Solution

Factor the denominator as

$$x^3 - x^2 - x + 1 = (x + 1)(x - 1)^2$$

Using Rules 1 and 2 for partial fractions,

$$\frac{2x + 1}{x^3 - x^2 - x + 1} = \frac{A}{x + 1} + \frac{B}{x - 1} + \frac{C}{(x - 1)^2}$$

for constants A, B, C. Cross multiplying to clear the denominators,

$$2x + 1 = A(x-1)^2 + B(x+1)(x-1) + C(x+1)$$

Setting $x = -1$,

$$-1 = 4A, \quad \text{so} \quad A = -\tfrac{1}{4}$$

Setting $x = 1$,

$$3 = 2C, \quad \text{so} \quad C = \tfrac{3}{2}$$

Setting $x = 0$,

$$1 = A - B + C, \quad \text{so} \quad B = \tfrac{1}{4}$$

So,

$$\begin{aligned}\int \frac{2x+1}{x^3 - x^2 - x + 1}\,dx &= -\frac{1}{4}\int \frac{dx}{x+1} + \frac{1}{4}\int \frac{dx}{x-1} + \frac{3}{2}\int \frac{dx}{(x-1)^2} \\ &= -\frac{1}{4}\ln|x+1| + \frac{1}{4}\ln|x-1| + \frac{3}{2}\left(-\frac{1}{x-1}\right) + C \\ &= \frac{1}{4}\ln\left|\frac{x-1}{x+1}\right| - \frac{3}{2(x-1)} + C\end{aligned}$$

6.2.3 Problems

1. Evaluate the following integrals:

(a) $\displaystyle\int \frac{dx}{x^2 - 9}$

(b) $\displaystyle\int \frac{x\,dx}{(x-5)^2}$

(c) $\displaystyle\int \frac{5x\,dx}{x^2 - 3x - 4}$

(d) $\displaystyle\int \frac{2x^2 + 3}{(x^2+1)^2}\,dx$

(e) $\displaystyle\int \frac{x^2 + x + 1}{x^2 - 2x - 8}\,dx$

6.3 VOLUMES OF REVOLUTION

When you revolve a plane area about a line, you generate a *solid of revolution*. The line is called the *axis of revolution*. See Figures 6.1 and 6.2.

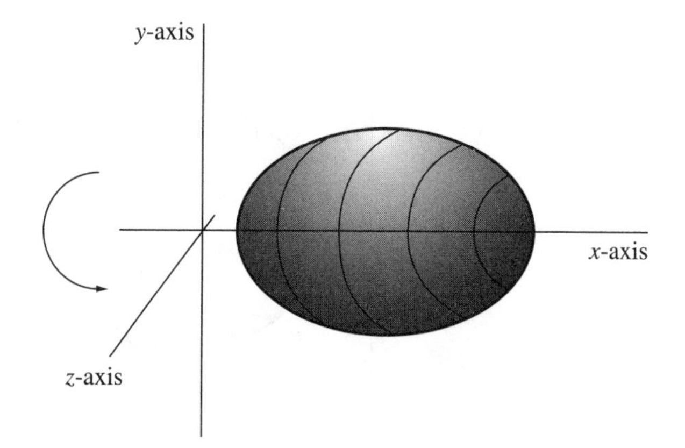

FIGURE 6.1 *Revolution about the x-axis*

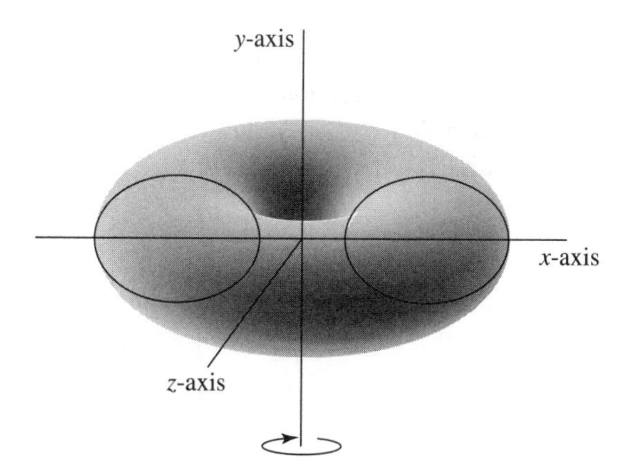

FIGURE 6.2 *Revolution about the y-axis*

There are three common methods for calculating the volume of a solid of revolution using integration: the disk method, the annulus method (sometimes called the 'washer method'), and the shell method. The choice of method used depends in part upon the shape of the solid.

6.3.1 The Disk Method

You can use the disk method to compute the volume of a solid of revolution when the axis of revolution is part of the boundary of the plane area.

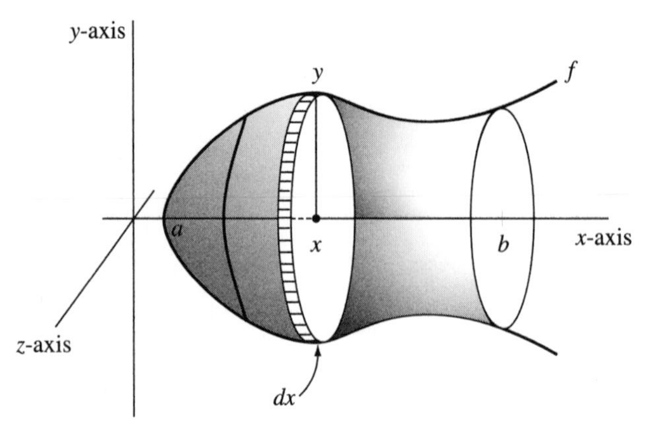

FIGURE 6.3 *The disk method*

Consider the case shown in Figure 6.3, where the plane area is bounded by the curve $y = f(x)$, the x-axis, and the line $x = b$.

The illustration shows a cross section of the solid at x, consisting of a thin disk of radius $y = f(x)$ and width dx. The volume of this disk is

$$\pi[f(x)]^2 dx$$

Since the disk at x has volume $\pi[f(x)]^2\,dx$, the volume of the entire solid is

$$V = \int_a^b \pi[f(x)]^2\,dx$$

Notice that the derivation just given uses the differential dx to indicate a small positive increment in the x-direction. A rigorous approach uses limits. [*Exercise*: Do this.]

6.3.2 The Annulus Method

You can use the annulus (or 'washer') method to compute the volume of a solid of revolution when the axis of revolution is not part of the boundary of the plane area.

In the case illustrated in Figure 6.4, the given plane area is bounded by the two curves $y = f(x)$, $y = g(x)$ and the lines $x = a$, $x = b$. In this case, the solid of revolution can be thought of as the solid generated by revolution of $y = f(x)$ with the inner solid generated by $y = g(x)$ removed from the center.

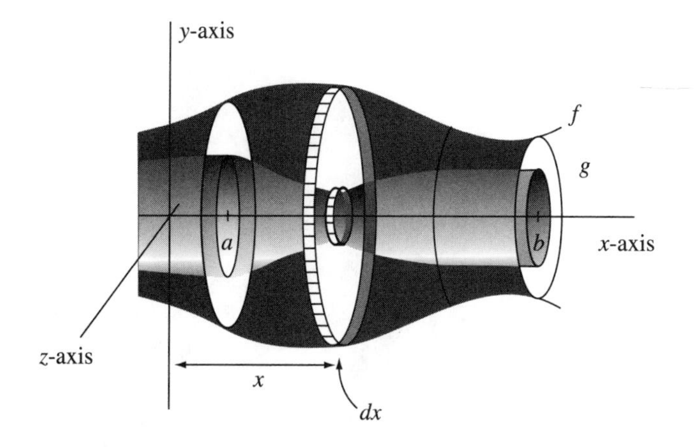

FIGURE 6.4 *The annulus method*

Consider a cross section of the solid at x, consisting of a thin annulus of width dx, outer radius $f(x)$, inner radius $g(x)$. The volume of this annulus is

$$\pi f(x)^2\,dx - \pi g(x)^2\,dx \;=\; \pi[f(x)^2 - g(x)^2]\,dx$$

Hence the volume of the solid of revolution is

$$\pi\int_a^b [f(x)^2 - g(x)^2]\,dx$$

6.3.3 The Shell Method

Volumes of solids formed by revolving a plane area about an axis can often be evaluated by the shell method.

Figure 6.5 (p. 168) shows an area revolved about the y-axis. Consider a narrow rectangular strip of the area at x, with width dx. Revolution of this strip around the y-axis yields a thin cylindrical shell of volume

$$2\pi x \times y \times dx = 2\pi x f(x) dx$$

The volume of the solid of revolution is the 'sum' of the volumes of all such shells, namely

$$2\pi\int_a^b x f(x)\,dx$$

6.3.4 Examples

✎ ***Example 1***

The region bounded by the curve $y = \sqrt{x}$, the x-axis, and the line $x = 4$ is rotated about the x-axis, generating a bullet-shaped region. Use the disk method to compute its volume.

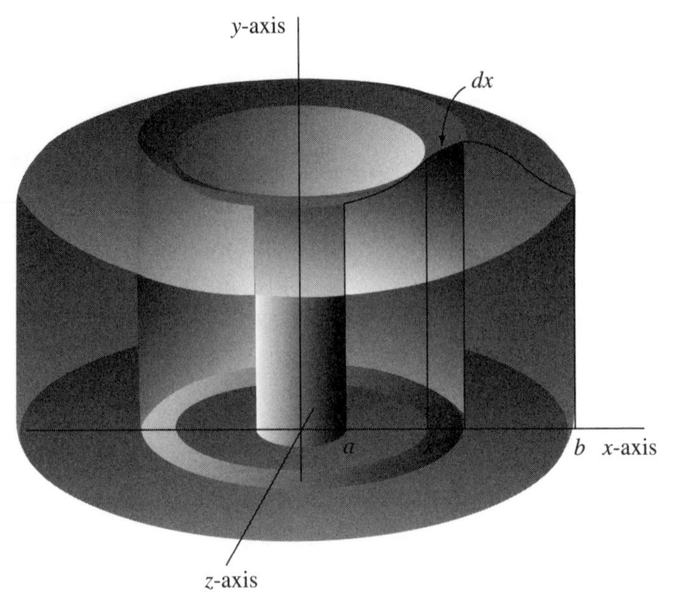

FIGURE 6.5 *The shell method*

Solution

It is a good idea to start out by drawing a sketch of the area and the solid it generates. I suggest you do that as an exercise.

By considering a thin disk of thickness dx at x, we get

$$V = \int_0^4 \pi \left(\sqrt{x}\right)^2 dx = \pi \int_0^4 x\,dx = \pi \left[\frac{x^2}{2}\right]_0^4 = 8\pi \text{ cubic units.}$$

✎ *Example 2*

The same region as in problem 1 is rotated about the y-axis to generate a solid of revolution. Use the annulus method to compute its volume.

Solution

As an exercise, you should draw a sketch of the solid generated. By considering a thin cross-section in the form of an annulus of thickness dy at y, we see that the volume of the solid is

$$\begin{aligned} \pi \int_0^2 [(4)^2 - (y^2)^2]\,dy &= \pi \int_0^2 (16 - y^4)\,dy \\ &= \pi \left[16y - \frac{y^5}{5}\right]_0^2 \\ &= \pi \left(32 - \frac{32}{5}\right) \end{aligned}$$

$$= \frac{128}{5}\pi \text{ cubic units}$$

✎ ***Example 3***

Use the shell method to compute the volume of the solid generated in problem 2.

Solution

By considering a thin section in the form of a cylindrical shell of thickness dx at x, this time we get the following expression for the volume:

$$\begin{aligned}
2\pi \int_0^4 x\sqrt{x}\,dx &= 2\pi \int_0^4 x^{3/2}\,dx \\
&= 2\pi \left[\frac{x^{5/2}}{5/2}\right]_0^4 \\
&= 2\pi \left[\frac{2x^{5/2}}{5}\right]_0^4 \\
&= 2\pi \frac{64}{5} \\
&= \frac{128}{5}\pi \text{ cubic units}
\end{aligned}$$

6.3.5 Problems

1. Draw a sketch of the solid of revolution generated by revolving about the x-axis the planar area enclosed by the curve $y = 2x^2$, the interval $[0, 5]$ of the x-axis, and the line $x = 5$. Use the disk method to compute the volume of this solid.

2. Take the planar area described in problem 1 and revolve it about the y-axis to generate a solid of revolution. Sketch the solid. Use the annulus method to compute the volume of this solid.

3. Use the shell method to compute the volume of the solid described in problem 2.

4. The planar area enclosed by one complete arch of the curve $y = \sin x$ above the x-axis is rotated about the x-axis to generate a solid of revolution. Compute its volume.

5. Draw the graph of the function $y = e^x \sin x$, for $0 \leq x \leq 3\pi$. The area enclosed by the first complete arch of this curve above the x-axis is rotated about the y-axis to generate a solid of revolution. Compute its volume. (See Examples 6.1.2, Example 6, p. 157.)

6.4 OTHER APPLICATIONS OF INTEGRATION

Integration has other applications in addition to the computation of areas and volumes. In this section we describe, in the briefest of ways, a few of those other applications.

6.4.1 Average Value

Given a finite set of numerical data points $x_0, \dots, x_{n-1}$, the average value of the data points is given by the expression

$$\frac{x_0 + \cdots + x_{n-1}}{n} = \frac{1}{n}\sum_{i=0}^{n-1} x_i$$

How can we define and calculate the average value of a continuous function $f(x)$ over an interval $[a, b]$?

If we take a finite subdivision $x_0 = a < x_1 < \cdots < x_n = b$ of the interval into points spaced an equal distance $\frac{b-a}{n}$ apart, then we can compute the average of the values of the function at the subdivision points $x_0, \dots, x_{n-1}$:

$$\frac{f(x_0) + \cdots + f(x_{n-1})}{n} = \frac{1}{n}\sum_{i=0}^{n-1} f(x_i)$$

If the subdivision points are sufficiently representative of the behavior of the function on $[a, b]$, that will give us an estimate of the average value of the function on $[a, b]$.

By taking more subdivision points, the estimate we obtain in this way will improve, since the finer the subdivision, the more representative will be the collection of those subdivision points. Thus, we can *define* the *average value* of the function $f(x)$ on $[a, b]$ to be the limit

$$\lim_{n\to\infty} \frac{1}{n}\sum_{i=0}^{n-1} f(x_i)$$

For each n, we have $h = \dfrac{b-a}{n}$, so we can rewrite this limit as

$$\lim_{n\to\infty} \frac{h}{b-a}\sum_{i=0}^{n-1} x_i = \frac{1}{b-a}\lim_{n\to\infty}\sum_{i=0}^{n-1} f(x_i)h$$

But the expression $\lim\limits_{n\to\infty}\sum\limits_{i=0}^{n-1} f(x_i)h$ on the right is just the definition of the definite integral $\displaystyle\int_a^b f(x)\,dx$. Hence,

$$\text{average value of } f(x) \text{ on } [a, b] = \frac{1}{b-a}\int_a^b f(x)\,dx$$

For example, the average value of the function $f(x) = x^2 + 1$ over the interval $[-1, 1]$ is

$$\begin{aligned}\frac{1}{2}\int_{-1}^{1}(x^2+1)\,dx &= \frac{1}{2}\left[\frac{x^3}{3}+x\right]_{-1}^{1} \\ &= \tfrac{1}{2}\left[(\tfrac{4}{3}) - (-\tfrac{4}{3})\right] \\ &= \tfrac{1}{2}\left[\tfrac{8}{3}\right] \\ &= \tfrac{4}{3}\end{aligned}$$

6.4.2 Work

In physics, we define the work done when a constant force F is used to move an object a distance d in the direction of the force to be equal to the product

$$W = F \times d$$

If F is measured in pounds and d is measured in feet, then W is given in ft-lbs. If F is measured in newtons and d is measured in meters, then W is given in joules (J). The conversion factor is roughly 1 ft-lb = 1.36 J.

For example, the work done in lifting an object weighing 20 lb a distance of 6 ft from the ground is

$$W = 20 \times 6 = 120 \text{ ft-lb}$$

How do we define and compute the work done when the force F varies continuously with the distance the object is moved?

Suppose that the force has the value $F = f(x)$ at point x, where the starting point is $x = a$ and where f is a continuous function. The variation in F in moving the object a very short distance dx from a point x to a point $x + dx$ will be small, so the work done in making this incremental move will be approximately

$$f(x)\,dx$$

The total work done in moving the object from the initial point a to a final point b will be approximated by the sum of all those quantities $f(x)\,dx$ as x goes from a to b. The smaller the increments dx, the better will be this approximation. Taking the limit as $dx \to 0$, we obtain the expression

$$W = \int_a^b f(x)\,dx$$

[*Exercise*: Fill in the details of the above discussion to go through the formal definition of the integral. Start by considering a subdivision $x_0 = a < x_1 < \cdots < x_n = b$ into equally spaced points and calculate an approximate value for the work done in moving the object from x_i to x_{i+1}.]

For example, Hooke's law says that the force exerted by a coiled spring stretched a distance x units beyond its unextended length is directly proportional to x:

$$F = kx$$

where k is a constant (called the *spring constant*). Hooke's law is only valid for values of x up to a certain point.

Suppose we take a spring of length 10 cm, and suppose that it takes a force of 40 newtons to maintain the spring extended to a length of 15 cm. The length x of extension is 5 cm or 0.05 meters, so Hooke's law gives

$$40 = 0.05k$$

and thus $k = 800$ N/m. The work done in stretching the spring from its natural position (10 cm) to 15 cm is

$$W = \int_0^{0.15} 800x\,dx = \left[400x^2\right]_0^{0.15} = 9 \text{ J}$$

6.4.3 Applications to Economics

Let's continue the discussion of economics started in Section 4.2.2. The *demand function* $p(x)$ is the price that a company has to charge in order to sell x units of a commodity. Usually, you need to lower prices in order to sell a greater quantity, so $p(x)$ is a decreasing function. Figure 6.6 shows a typical demand function.

If X is the amount of the commodity currently available, then $P = p(X)$ is the current selling price.

Suppose we partition the interval $[0, X]$ into n equally sized subintervals, each of length $h = X/n$, with subdivision points $x_0 = 0 < x_1 < \cdots < x_n = X$. If h is small, we can regard the price as fixed throughout the interval $(x_i, x_{i+1}]$ at the lowest actual value $p(x_{i+1})$ in that interval. This gives us a modified demand function in the form of a step function.

[*Exercise*: Draw a graph to illustrate this. Start with the graph shown in Figure 6.6, add the subdivision points, along with verticals from those points up to the demand curve, and then draw in the constant price lines over each subdivision. Already you can see where integration might creep in.]

Taking our modified demand function, let's look at the customers represented in the interval $[x_{i-1}, x_i]$. They are the ones who will buy when the price falls to $p(x_i)$, but will not buy when it is any higher. In buying at the price P, they are saving an amount $p(x_i) - P$ per unit

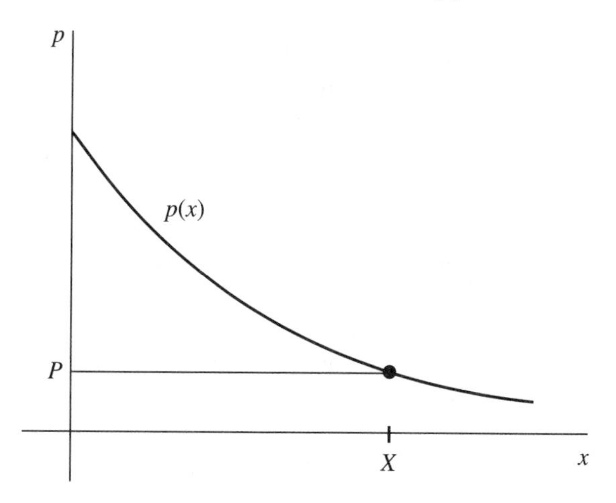

FIGURE 6.6 *The demand curve*

on the amount they would be prepared to pay. (They are reaping the benefit of the increased sales resulting from the entry of others into the market.) So the total amount saved for these customers is

$$[p(x_i) - P]h$$

The total amount saved by all the customers who buy at the current price P is thus

$$\sum_{i=1}^{n}[p(x_i) - P]\,h$$

Taking the limit as $n \to \infty$, this sum approaches the integral

$$\int_0^X [p(x) - P]\,dx$$

This integral is what economists call the *consumer's surplus* for the commodity. It represents the amount of money saved by customers in purchasing the commodity at price P corresponding to a demand level of X. Diagramatically, the consumer's surplus is represented by the area between the demand curve $p(x)$ and the line $p = P$ in Figure 6.6.

6.4.4 Application in Human Anatomy

We continue the discussion on blood flow begun in Section 4.2.3. Our starting point is Poiseuille's equation for laminar flow:

$$v(r) = \frac{P}{4\eta L}(R^2 - r^2)$$

that gives the velocity $v(r)$ of the blood flowing a distance r from the central axis of a blood vessel of radius R and length L, where P is the pressure difference between the ends of the vessel and η is the viscosity of the blood.

We are interested in knowing the volume of blood that flows in unit time, what is called the *flux*. To compute the flux, we take a cross-section of the vessel and subdivide the cross-section into concentric annuli. To do this, we subdivide the interval $[0, R]$ into equally sized subintervals, by means of subdivision points $r_0 = 0 < r_1 < \cdots < r_n = R$, spaced a distance $h = R/n$ apart.

The area of the annulus with inner radius r_i and outer radius r_{i+1} is approximately equal to

$$2\pi r_{i+1} h$$

If h is small, the velocity of the blood flowing across this annulus can be regarded as constant and approximated by $v(r_{i+1})$. Thus the volume of blood per unit time that flows across the annulus is approximately

$$2\pi r_{i+1} h v(r_{i+1})$$

Hence, the total volume of blood that flows across the entire cross-section of the vessel in unit time is approximately

$$\sum_{i=0}^{n-1} 2\pi r_{i+1} h v(r_{i+1})$$

The finer the subdivision (i.e., the smaller is h and the larger is n), the more accurate this approximation will be. Taking the limit as $h \to 0$, we obtain the following integral for the exact value of the flux:

$$F = \int_0^R 2\pi r v(r)\, dr$$

Let's evaluate this integral:

$$\begin{aligned} F &= \int_0^R 2\pi r v(r)\, dr \\ &= \int_0^R 2\pi r \frac{P}{4\eta L}(R^2 - r^2)\, dr \\ &= \frac{\pi P}{2\eta L} \int_0^R (R^2 r - r^3)\, dr \\ &= \frac{\pi P}{2\eta L} \left[\frac{R^2 r^2}{2} - \frac{r^4}{4} \right]_0^R \\ &= \frac{\pi P}{2\eta L} \left[\frac{R^4}{2} - \frac{R^4}{4} \right] \\ &= \frac{\pi P R^4}{8\eta L} \end{aligned}$$

This expression for F is called *Poiseuille's law*. It shows that the flux is proportional to the fourth power of the radius of the blood vessel. It explains, in dramatic fashion, why high blood cholesterol is such a killer. As fat deposits build up on the lining of a blood vessel, the effective radius of the vessel is decreased. By the above equation, the blood flow decreases with the fourth power of the radius. For example, halving the radius produces a sixteenfold reduction in blood flow! Perhaps the key to motivating people to change their diet and take up exercise is to encourage them to take a calculus course!

6.4.5 Problems

1. Compute the average value of the function $\sin x$ over the interval $[0, \pi]$.

2. A weightlifter lifts a 60-kg barbell from the floor to a height of 2.4 m. Calculate the amount of work he does in making the lift.

3. A spring has a natural length of 20 cm. A 25-newton force will keep the spring stretched to a length of 30 cm. Compute the amount of work required to stretch the spring from 20 cm to 30 cm.

4. When a gas (usually a mixture of air and gasoline vapor) expands in a cylinder of an automobile engine, the pressure at any given time is a function of the volume: $P = P(V)$. The force exerted by the gas on the piston is the product of the pressure and the area of the piston face. If r is the radius of the cylinder, then the force is

$$F = \pi r^2 P$$

Show that the work done by the gas when the volume expands from V_1 to V_2 is

$$\int_{V_1}^{V_2} P\,dV$$

5. A demand curve is given by

$$p = \frac{1000}{x + 30}$$

Find the consumers' surplus when the selling price is \$20.

6. Using the equation for laminar flow, find the average velocity of the flow of blood in an artery over the interval $0 \leq r \leq R$.

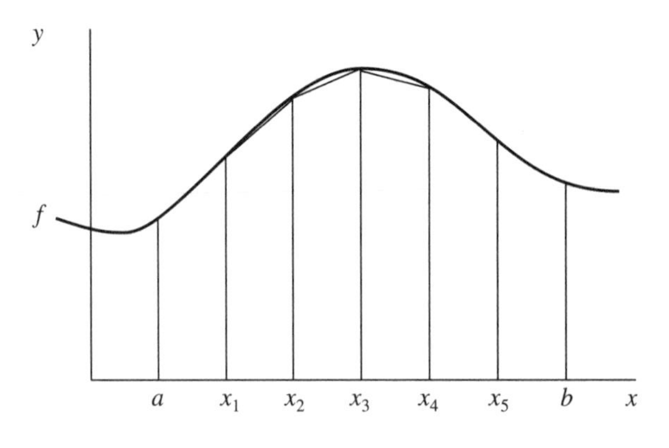

FIGURE 6.7 *The Trapezoidal Rule*

6.5 NUMERICAL INTEGRATION

For many functions, you cannot evaluate an antiderivative, or it is very difficult to do so. If you need a definite integral of such a function, you have to use a numerical method. The topic of numerical integration is part of a large subject known as *numerical methods*, most of which is beyond the scope of the *Electronic Companion*.

In general, any numerical method will produce only an approximate answer, so its use should be accompanied by an analysis of the likely error.

6.5.1 The Trapezoidal Rule

The Trapezoidal Rule lets you find an approximation to a definite integral. It works by approximating the area under the curve $y = f(x)$ between $x = a$ and $x = b$ by a sequence of trapezoids. The approximation depends on the choice of a subdivision

$$a = x_0 < \cdots < x_n = b$$

of the interval $[a, b]$ into n equal subintervals of length $h = (b - a)/n$. Given such a subdivision, the Trapezoidal Rule gives the approximation

$$\int_a^b f(x)\,dx \approx \frac{h}{2}[f(x_0) + 2f(x_1) + \cdots + 2f(x_{n-1}) + f(x_n)]$$

In general, the larger the value of n, the better will be the approximation.

Here is the derivation of the above formula. For each subdivision point x_i, draw the vertical line to the curve. The illustration in Figure 6.7 shows the case $n = 6$.

Join the adjacent subdivision points on the curve by chords, thereby approximating the curve by a sequence of straight line segments. The trapezoid determined by the subinterval $[x_i, x_{i+1}]$ has area

$$\tfrac{1}{2} \times [f(x_i) + f(x_{i+1})] \times h$$

The total area of the trapezoids is:

$$\tfrac{1}{2}[f(x_0) + f(x_1)]\,h + \tfrac{1}{2}[f(x_1) + f(x_2)]\,h + \cdots$$
$$\cdots + \tfrac{1}{2}[f(x_{n-2}) + f(x_{n-1})]\,h + \tfrac{1}{2}[f(x_{n-1}) + f(x_n)]\,h$$

Collecting terms, this sum works out to be

$$\frac{h}{2}[f(x_0) + 2f(x_1) + \cdots + 2f(x_{n-1}) + f(x_n)]$$

This is our approximation to the integral.

For example, let's use the Trapezoidal Rule to approximate the integral

$$\int_0^{1/2} \frac{dx}{1+x^2}$$

We'll take $n = 5$. Then $h = \dfrac{1/2 - 0}{5} = 0.1$ and the interval subdivision points are 0.1, 0.2, 0.3, 0.4. Hence

$$\begin{aligned}\int_0^{1/2} \frac{dx}{1+x^2} &\approx \frac{0.1}{2}\big[f(0) + 2f(0.1) + 2f(0.2) + 2f(0.3) + 2f(0.4) + f(0.5)\big] \\ &= \frac{1}{20}\left[1 + \frac{2}{1.01} + \frac{2}{1.04} + \frac{2}{1.09} + \frac{2}{1.16} + \frac{1}{1.25}\right] \\ &= 0.4631\end{aligned}$$

For comparison, direct integration gives

$$\int_0^{1/2} \frac{dx}{1+x^2} = \big[\arctan x\big]_0^{1/2} = 0.4636$$

so the approximation given by the Trapezoidal Rule with $n = 5$ is accurate to three decimal places.

6.5.2 Simpson's Rule

Simpson's Rule provides a common alternative to the Trapezoidal Rule for computing a numerical approximation to a definite integral. Whereas the Trapezoidal Rule approximates the curve by straight-line segments that meet the given curve at their two endpoints, Simpson's Rule uses short lengths of quadratic curves that meet the given curve at their endpoints and their midpoints. In general, Simpson's Rule gives a much better approximation to the integral than does the Trapezoidal Rule.

The Simpson's Rule formula is:

$$\int_a^b f(x)\,dx \approx \frac{h}{3}[f(x_0) + 4f(x_1) + 2f(x_2) + 4f(x_3) + 2f(x_4) + \cdots$$
$$\cdots + 2f(x_{n-2}) + 4f(x_{n-1}) + f(x_n)]$$

Notice the pattern of the coefficients: $1, 4, 2, 4, 2, 4, 2, \ldots, 2, 4, 1$. (For this to work, n must be odd.) The formula results from simplification of the formula

$$\frac{h}{3}[(f(x_0) + 4f(x_1) + f(x_2)) + (f(x_2) + 4f(x_3) + f(x_4)) + \cdots$$
$$\cdots + (f(x_{n-2}) + 4f(x_{n-1}) + f(x_n))]$$

where the sum consists of triples of the form

$$(f(x_{2i}) + 4f(x_{2i+1}) + f(x_{2i+2}))$$

Each such term results from approximating the section of the curve $y = f(x)$ between x_{2i} and $x_{2(i+1)}$ by a quadratic $y = Ax^2 + Bx + C$ that meets the curve at $x = x_{2i},\ x_{2i+1},\ x_{2(i+1)}$. To see how this happens, let's start by looking at the case $n = 2$.

Let $f(x)$ be given. Consider a quadratic curve

$$Ax^2 + Bx + C$$

that meets the curve $y = f(x)$ at $x = a, b, c$, where

$$c = \frac{a+b}{2}$$

See Figure 6.8. The area beneath the quadratic between $x = a$ and $x = b$ is:

$$\begin{aligned}
\int_a^b [Ax^2 + Bx + C]\,dx &= \left[\frac{Ax^3}{3} + \frac{Bx^2}{2} + Cx\right]_a^b \\
(1) \qquad &= \frac{A}{3}(b^3 - a^3) + \frac{B}{2}(b^2 - a^2) + C(b-a) \\
&= \frac{b-a}{3}\left[A(a^2+ab+b^2) + \frac{3B}{2}(a+b) + 3C\right]
\end{aligned}$$

Now, since the quadratic meets the curve $y = f(x)$ at $x = a, b, c$, we have

$$\begin{aligned}
Aa^2 + Ba + C &= f(a) \\
Ab^2 + Bb + C &= f(b) \\
Ac^2 + Bc + C &= f(c)
\end{aligned}$$

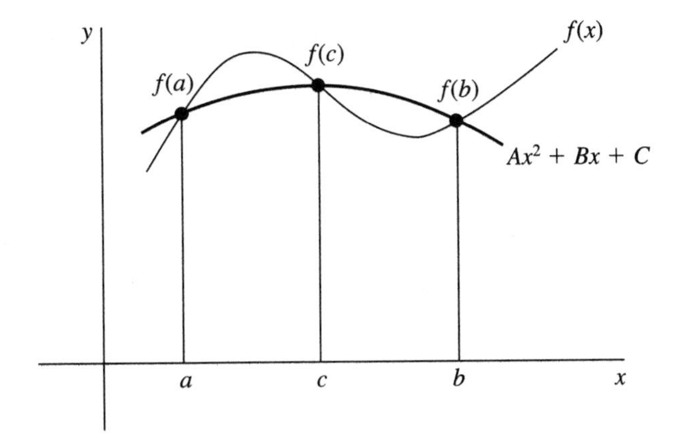

FIGURE 6.8 *Simpson's Rule*

Thus,

$$
\begin{aligned}
f(a) &+ 4f(c) + f(b) \\
&= Aa^2 + Ba + C \\
&\qquad + 4\left[A\left(\frac{a+b}{2}\right)^2 + B\left(\frac{a+b}{2}\right) + C\right] + Ab^2 + Bb + C \\
&= A(a^2+b^2) + B(a+b) + 2C + A(a^2+2ab+b^2) + 2B(a+b) + 4C \\
&= 2A(a^2+ab+b^2) + 3B(a+b) + 6C
\end{aligned}
$$

Hence,

$$\text{(2)} \quad \frac{1}{2}[f(a) + 4f(c) + f(b)] = A(a^2+ab+b^2) + \frac{3B}{2}(a+b) + 3C$$

Hence, by Eqs. (1) and (2),

$$\int_a^b [Ax^2 + Bx + C]\,dx = \frac{b-a}{6}[f(a) + 4f(c) + f(b)]$$

If we set $h = \dfrac{b-a}{2}$, this becomes

$$\text{(3)} \qquad \int_a^b [Ax^2 + Bx + C]\,dx = \frac{h}{3}[f(a) + 4f(c) + f(b)]$$

To obtain Simpson's formula, subdivide the interval $[a, b]$ into $n = 2m$ equally sized subintervals

$$x_0 = a < x_1 < x_2 < \cdots < x_{n-2} < x_{n-1} < x_n = b$$

and apply Eq. (3) with $a = x_{2i}$, $b = x_{2(i+1)}$, $c = x_{2i+1}$, for each index $i = 1, 2, \ldots, m$, in turn.

For an example of Simpson's Rule, let's use it to approximate the same integral we used to illustrate the Trapezoidal Rule, namely

$$\int_0^{1/2} \frac{dx}{1+x^2}$$

We'll take $n = 4$. (For the Trapezoidal Rule we took $n = 5$. The number of subdivisions has to be even for Simpson's Rule.) Then $h = \dfrac{1/2 - 0}{4} = \dfrac{1}{8}$ and the internal subdivision points are $\frac{1}{8}$, $\frac{1}{4}$, $\frac{3}{8}$. Hence

$$\begin{aligned}
&\int_0^{1/2} \frac{dx}{1+x^2} \\
&\quad \approx \frac{1}{24}\left[1 + 4\frac{1}{1+(\frac{1}{8})^2} + 2\frac{1}{1+(\frac{1}{4})^2} + 4\frac{1}{1+(\frac{3}{8})^2} + \frac{1}{1+(\frac{1}{2})^2}\right] \\
&\quad = \frac{1}{24}\left[1 + \frac{256}{65} + \frac{32}{17} + \frac{256}{73} + \frac{4}{5}\right] \\
&\quad = 0.46365
\end{aligned}$$

We saw earlier that integration gives

$$\int_0^{1/2} \frac{dx}{1+x^2} = \big[\arctan x\big]_0^{1/2} = 0.46365$$

so the approximation given by Simpson's Rule with $n = 4$ is accurate to five decimal places. This is already two more places of accuracy than the Trapezoidal Rule with $n = 5$.

6.5.3 Accuracy

The accuracy of the Trapezoidal Rule and Simpson's Rule depends on how reasonable it is to approximate the given curve by chords in the first place and quadratics in the second.

For the Trapezoidal Rule, if the curve is concave up throughout the interval of integration, the approximation will be too high, since all the chords will lie above the curve; if the curve is concave down throughout the interval, the approximation will be too low. If the concavity changes over the interval, approximation errors can cancel each other out to some extent.

In general, the error in the Trapezoidal Rule is roughly proportional to $1/n^2$. Thus, doubling n increases the accuracy by a factor of 4. For fixed n, the size of the error depends on the size of the second derivative f''.

The error in Simpson's Rule is roughly proportional to $1/n^4$. Thus, doubling n increases the accuracy by a factor of 16. For fixed n, the size of the error depends on the size of the fourth derivative $f^{(4)}$.

In general then, Simpson's Rule is much more accurate than the Trapezoidal Rule.

6.5.4 Problems

1. Using a programmable calculator or a computer, write a routine to evaluate $\int_1^5 \frac{dx}{x}$ using the Trapezoidal Rule with $n = 5,\ 11,\ 21,\ 101$, and compare the answers you get.

2. Using a programmable calculator or a computer, write a routine to evaluate $\int_1^5 \frac{dx}{x}$ using Simpson's Rule with $n = 5,\ 11,\ 21,\ 101$, and compare the answers you get.

   For each n, compare your answer with the corresponding answer in problem 1.

3. Using a programmable calculator or a computer, evaluate $\int_1^2 \ln x\, dx$, using (a) the Trapezoidal Rule with $n = 5,\ 11,\ 101$ and (b) Simpson's Rule with the same values of n.

   Compare the different answers you get. Then use integration by parts to determine the integral exactly.

4. The following table gives the values of a quantity q read off from a measuring device at one-minute intervals over a period of 10 minutes:

t	0	1	2	3	4	5	6	7	8	9	10
q	0	1.1	1.6	1.8	2.5	2.7	2.8	2.6	2.0	1.2	0.3

   You know that q varies continuously with t, but you have no idea what function relates t and q, or even if the function can be expressed using standard mathematical formulas.

   Use (a) the Trapezoidal Rule and then (b) Simpson's Rule to estimate

$$\int_0^{10} q\, dt$$

   Under what circumstances, if any, will the estimate in (a) be more reliable than that in (b)?

   When, if ever, will (b) be the more reliable estimate?

   Suppose you are able to find a polynomial approximation $p(x)$ for the unknown function $q(t)$ that gives exactly the values in the table for $t = 0,\ 1,\ 2,\ 3,\ 4,\ 5,\ 6,\ 7,\ 8,\ 9,\ 10$. Which would be more likely to give the better estimate of $\int_0^{10} q\, dt$, the estimate obtained by Simpson's Rule or the estimate given by $\int_0^{10} p(t)\, dt$? Investigate and discuss.

Answers to Problems

Only answers to problems with specific solutions are provided.

1.2.5

1. **(a)** $(-\infty, -\frac{1}{5}], [\frac{3}{5}, \infty)$ **(b)** $(-\infty, -1), (2, \infty)$

   (c) $(-3, -2), (-1, \infty)$ **(d)** $(-1, \infty)$ **(e)** $(-\infty, -3),\ (-3, -\frac{11}{6})$

1.3.4

1. No
2. Yes
3. $(\frac{3}{2}, \frac{9}{2})$
4. $y = \frac{19}{14}x - \frac{48}{7}$
5. $y = \sqrt{3}x + 7$

1.4.3

1. **(a)** $(-\infty, -3], [3, \infty)$ **(b)** $[-1, 1]$ **(c)** $(-1, 1)$

   (d) $(-\infty, 1), (1, \infty)$ **(e)** $(-\infty, \infty)$

2. **(a)** $g(x) = \frac{1}{6}x - \frac{11}{6}$ **(b)** $g(x) = \dfrac{x}{1-x}$, domain $= [0, 1)$

1.5.5

1. $\sin(21\pi) = 0$; $\cos(-11\pi) = -1$; $\tan(9\pi) = 0$;

   $\sin(\frac{13\pi}{6}) = \frac{1}{2}$; $\cos(\frac{19\pi}{3}) = \frac{1}{2}$; $\tan(\frac{9\pi}{4}) = 1$

1.6.4

1. 2; 3; $\frac{1}{2}$; $x + 1$; $\log_e(x + 1) - 2$

2.1.5

1. **(a)** $\frac{1}{2}, \frac{1}{6}, \frac{1}{12}, \frac{1}{20}$ **(b)** $2, 1, \frac{8}{9}, 1$ **(c)** $-\frac{1}{2}, \frac{2}{3}, -\frac{3}{4}, \frac{4}{5}$

2. **(a)** $\frac{n}{n+1}$ **(b)** $\frac{(-1)^{n+1}}{2n+1}$ **(c)** $\frac{n(n+1)}{(n+2)(n+3)}$

3. **1(a)** 0 **1(b)** ∞ **1(c)** no limit **2(a)** 1 **2(b)** 0 **2(c)** 1

2.2.8

1. **(a)** -17 **(b)** 1 **(c)** $\frac{1}{6}$ **(d)** 0 **(e)** no limit

2. limit equals 0

4. **(a)** 2 **(b)** $\frac{3}{4}$ **(c)** 0

2.3.5

1. **(a)** none **(b)** $x = -1$ **(c)** none **(d)** none **(e)** $x = \pm 2$

(f) $x = 2, 3$ **(g)** $x = 2, 3$

2. **(b)** not possible **(e)** can redefine $f(2) = 0$

(g) can redefine $f(2) = -4$

3.1.5

1. **(a)** $5x^4 + 66x^2 - 2x + 15$ **(b)** $\frac{64}{7}x^{-3/7}$ **(c)** $2\pi(\pi+1)x^{2\pi-1}$

(d) $-\frac{11}{4}x^{-15/4}$ **(e)** $-\frac{1}{x^2}$ **(f)** $\frac{-15.705}{x^6}$ **(g)** $-\frac{1}{2x^{3/2}}$

(h) $\log_5 9 \log_2 5\, x^{\log_2 5 - 1}$ **(i)** 0

2. The value of the derivative is $f'(3) = -1$.

3.2.6

1. **(a)** $9(3x^3 + 2x^2 + x + 1)^8(9x^2 + 4x + 1)$

(b) $\frac{4(x-1)}{(x+1)^3}$

(c) $\frac{1}{(x+1)^2}\sqrt{\frac{x+1}{x-1}}$

(d) $(x^3 + x^2 + x + 1)(5x^4 + 4x^3) + (x^5 + x^4 + 1)(3x^2 + 2x + 1)$

(e) $\frac{(x^5 + x^3 + x + 1)(6x^5 + 4x^3 + 2x) - (x^6 + x^4 + x^2)(5x^4 + 3x^2 + 1)}{(x^5 + x^3 + x + 1)^2}$

(f) $\sqrt{1-x^2} - \dfrac{x^2}{\sqrt{1-x^2}} = \dfrac{1-2x^2}{\sqrt{1-x^2}}$

(g) $\dfrac{\sqrt{x^2+1} - x^2/\sqrt{x^2+1}}{x^2+1} = \dfrac{1}{(x^2+1)^{3/2}}$

2. $2x - 6$

3. $\dfrac{\sqrt{1-y^2}}{1-2y^2}$

4. $-\frac{1}{9}$

3.3.2

1. **(a)** $-15 \csc 3x \cot 3x - 14 \sec 7x \tan 7x$

(b) $5 \sec^4 x \sec x \tan x = 5 \sec^5 x \tan x$

(c) $\dfrac{-\csc^2 x}{2\sqrt{\cot x}} = -\dfrac{1}{2} \csc^2 x \sqrt{\tan x}$

(d) $\sec x \sec^2 x + \sec x \tan x \tan x = \sec^3 x + \sec x \tan^2 x$

(e) $\dfrac{\cos\sqrt{x}}{2\sqrt{x}}$ **(f)** $12x^3 \sec^2(3x^4)$ **(g)** $\dfrac{(1+\tan x) - x \sec^2 x}{(1+\tan x)^2}$

3.4.2

1. **(a)** $35e^{7x}$ **(b)** $2xe^{x^2}$ **(c)** $\dfrac{15x^2}{5x^3} = \dfrac{3}{x}$ **(d)** $\pi x^2 e^{\pi x - 1} + 2xe^{\pi x}$

(e) $\dfrac{-4}{(e^x - e^{-x})^2}$ (after simplification) **(f)** $\dfrac{e^{\sqrt{x}}}{2\sqrt{x}} + \dfrac{1}{2x}$

(g) $e^x(\cos x - \sin x) + e^x(\sin x + \cos x) = 2e^x \cos x$

(h) $(\cos x)e^{\sin x}$ **(i)** $\dfrac{2x^2}{1+x^2} + \ln(1+x^2)$ **(j)** $\ln x$

3.5.3

1. $f^{(n)}(x) = e^x + (-1)^n e^{-x}$

2. $f^{(n)}(x) = \begin{cases} 4x^3, & n = 1 \\ 12x^2, & n = 2 \\ 24x, & n = 3 \\ 24, & n = 4 \\ 0, & n \geq 5 \end{cases}$

3. $f^{(n)}(x) = \begin{cases} (-1)^{(n-1)/2}\, 2^n \cos 2x, & \text{if } n \text{ is odd,} \\ (-1)^{n/2}\, 2^n \sin 2x, & \text{if } n \text{ is even.} \end{cases}$

3.6.3

1. $x_0 = \sqrt{\dfrac{13}{3}} \approx 2.08$

4.1.7

1. Critical numbers $x = -2, -\frac{1}{2}, 1$; local minimum at $(-2, 0)$; local maximum at $(-\frac{1}{2}, \frac{81}{16})$; local minimum at $(1, 0)$. Second derivative vanishes for $x = \dfrac{\pm\sqrt{5} - 1}{2} \approx -1.618, 0.618$. For $x < -1.618\ldots$, the curve is concave up; for $-1.618\ldots < x < 0.618\ldots$, the curve is concave down; for $x > 0.618\ldots$, the curve is concave up.

2. Critical number $x = 2$; local minimum at $(2, 0)$ (a cusp); concave down everywhere.

3. Critical numbers $x = -2, 0, 1$; local maximum at $(-2, 0)$ (cusp); local minimum at $(0, -\sqrt[3]{4})$; point of inflection at $(1, 0)$ (slope infinite); concave up for $x < 1$; concave down for $x > 1$.

4. Critical numbers $x = 0, \frac{1}{2}, 1$.

4.2.6

1. $v = 3t^2 - 10t + 8.5$, $a = 6t - 10$; $v = 0$ has no real roots, so never at rest; v is a minimum at $t = \frac{5}{4}$ secs; v is never a maximum according to the equation given.

2. $q = 2,500$

3. $P(t) = 100,000(1.05)^t$, where $t = 0$ is the year 1990. In 2001, $t = 11$ and $P(11) = 171,034$. In 2050, $t = 60$ and $P(60) = 1,867,919$

4. $M(t) = 300e^{-kt}$, where $k = \dfrac{\ln 2}{140} \approx 0.00495$. For $t = 100$, $M(100) = 182.87$ mg. For $M(t) = 20$, $t = 547.08$ days.

5. $M(t) = M_0 e^{-kt}$. $M(3) = 0.42M_0$, so $k \approx 0.28917$. Half-life $T = \dfrac{\ln 2}{k} \approx 2.39702$ days. For $M(t) = 0.1M_0$, $t = 7.96$ days.

4.3.2

1. Approximately 81.8 mph

2. Approximately 1.677 m/sec

4.4.2

1. (a) $\frac{dy}{dx} = \frac{4y - 3x^2}{3y^2 - 4x}$ (b) $\frac{dy}{dx} = \frac{2x \sin y + \sin(x - y)}{\sin(x - y) - x^2 \cos y}$

   (c) $\frac{dy}{dx} = \left[2 - \sqrt{y+1} - \frac{y}{2\sqrt{1+x}}\right] \Big/ \left[\frac{x}{2\sqrt{1+y}} + \sqrt{1+x}\right]$

4.5.4

1. (a) $dy = -\frac{1}{2\sqrt{1-x}}dx$; at $x = 0$, $dy = -\frac{1}{2}\,dx$; for $dx = 0.01$, $dy = -0.005$

   (b) $dy = \cos x\,dx$; at $x = \frac{\pi}{6}$, $dy = \frac{\sqrt{3}}{2}\,dx$; for $dx = 0.5$, $dy = \frac{\sqrt{3}}{4} \approx 0.433$

   (c) $dy = \frac{2}{(x+1)^2}dx$; at $x = 1$, $dy = \frac{1}{2}\,dx$; for $dx = -0.1$, $dy = -0.05$

2. (a) $6\frac{1}{54} \approx 6.0185185$ (b) $9\frac{19}{20} = 9.95$ (c) $\frac{\sqrt{3}}{2} - \frac{\pi}{360} \approx 0.857298758$

   The calculator answers are: $\sqrt[3]{218} \approx 6.01845$, $\sqrt{99} \approx 9.94987$, $\sin(59^\circ) \approx 0.85717$.

4.6.2

1. (a) $\frac{dy}{dx} = \frac{\sin u}{2(1 - \cos u)}$ (b) $\frac{dy}{dx} = \csc u$ (after simplifying)

   (c) $\frac{dy}{dx} = \frac{t^2 - 1}{2t}$ (after simplifying)

2. (a) not easily evaluated (b) $x^2 - y^2 = 1$ (c) $x^2 + y^2 = 1$

4.7.3

1. (a) $\frac{2}{\sqrt{1 - 4x^2}}$ (b) $\frac{2 \arcsin x}{\sqrt{1 - x^2}}$ (c) $\frac{\arccos x}{x} - \frac{\ln x}{\sqrt{1 - x^2}}$

   (d) $\frac{1 - 2x \arctan x}{(1 + x^2)^2}$ (e) $\frac{1}{x\sqrt{x^2 - 1}}$

5.2.3

1. **(a)** 4 **(b)** 1 **(c)** $e-1$ **(d)** 50 **(e)** 1

5.3.3

1. **(a)** $\dfrac{x^{12}}{12}-\dfrac{x^9}{3}+7x^3-5x+C$ **(b)** $\dfrac{7^x}{\ln 7}+C$

   (c) $-\frac{11}{3}\cos 3x-\frac{7}{5}\sin 5x+C$ **(d)** $-\frac{1}{\pi}\cos \pi x+C$ **(e)** $ex+C$

   (f) $\dfrac{e^{8.2x}}{8.2}+C$ **(g)** $\dfrac{x^2}{2}-\dfrac{1}{x}+C$ **(h)** $\frac{2}{3}x^{3/2}-\frac{2}{5}x^{5/2}+C$

   (i) $3\ln x+C$ **(j)** $-\dfrac{1}{4x^4}+C$

5.4.2

Note that there are several different ways to express each of the answers to this problem.

1. **(a)** $\frac{1}{2}x-\frac{1}{4}\sin 2x+C$ **(b)** $-\frac{2}{3}\cos^3 x-\sin^2 x\cos x+C$

   (c) $\frac{1}{8}x-\frac{1}{32}\sin 4x+C$

   (d) $\frac{3}{8}x-\frac{3}{8}\sin x\cos x-\frac{1}{4}\sin^3 x\cos x+C$

   (e) $-\cos x+\frac{2}{3}\cos^3 x-\frac{1}{5}\cos^2 x+C$

5.5.1

1. **(a)** $-\frac{14}{3}$ **(b)** π **(c)** $\frac{1}{2}(e^8-1)$ **(d)** $\frac{2}{3}(2)^{3/2}$ **(e)** $\ln 3$

5.6.5

1. **(a)** $\dfrac{(x+1)^{13}}{13}+C\ (u=x+1)$ **(b)** $\frac{2}{3}\sqrt{x^3+1}+C\ (u=x^3+1)$

   (c) $-e^{1/x}+C\ \left(u=\dfrac{1}{x}\right)$

   (d) $\frac{1}{2}x\sqrt{x^2-25}+\frac{25}{2}\ln|x+\sqrt{x^2-25}|+C\ (x=5\sec z)$

   (e) $\dfrac{-\sqrt{3+x^2}}{3x}+C\ (x=\sqrt{3}\tan z)$

   (f) $\dfrac{1}{2}\ln\left|\dfrac{\sqrt{4+9x^2}-2}{x}\right|+C\ (x=\frac{2}{3}\tan z)$

(g) $\sqrt{9x^2+4}+\frac{2}{3}\ln\left|\frac{3\sqrt{9x^2+4}-4}{3\sqrt{9x^2+4}+4}\right|+C \ (x=\frac{2}{3}\tan z)$

(h) $\frac{x}{\sqrt{1-x^2}}+C \ (x=\sin z)$ (i) $\ln|\sin x|+C \ (u=\sin x)$

2. (a) $\frac{1603}{9}$ (b) $\frac{1}{2}\ln 3$ (c) $\frac{\sqrt{3}}{2}$

5.7.2

1. $\frac{37}{12}$

6.1.3

1. (a) $\frac{2}{3}x(1+x)^{3/2}-\frac{4}{15}(1+x)^{5/2}+C$ (b) $(x^2-2x+2)e^x+C$

   (c) $(\frac{1}{2}x^3-\frac{3}{4}x^2+\frac{3}{4}x-\frac{3}{8})e^{2x}+C$

   (d) $x\ln(1+x^2)-2x+2\arctan x+C$

   (e) $\frac{1}{2}(\sec x\tan x+\ln|\sec x+\tan x|)+C$

2. (a) $x\arcsin x+\sqrt{1-x^2}+C$ (b) $x\arccos x-\sqrt{1-x^2}+C$

   (c) $x\arctan x-\ln\sqrt{1+x^2}+C$

6.2.3

1. (a) $\frac{1}{6}\ln\left|\frac{x-3}{x+3}\right|+C$ (b) $\ln|x-5|-\frac{5}{x-5}+C$

   (c) $\ln|(x+1)(x-4)^4|+C$ (d) $\frac{5}{2}\arctan x+\frac{x}{2(x^2+1)}+C$

   (e) $x-\frac{1}{2}\ln|x+2|+\frac{7}{2}\ln|x-4|+C$

6.3.5

1. $2,500\pi$
2. 625π
3. 625π
4. $\pi^2/2$
5. $\pi[(\pi-1)e^\pi-1]$

6.4.5

1. $2/\pi$
2. 144 kg-m
3. 12.5 J
5. \$110.83
6. $\dfrac{PR^2}{6\eta L}$

6.5.4

1. Integration gives the exact answer 1.60944.
3. Integration gives the exact answer 0.38629.
4. **(a)** 18.45 **(b)** 18.56

index